家族荣耀
豪宅铭刻

Family Glory by
Individual Mansion

下

open 欧朋文化 策划

黄滢 马勇 主编

华中科技大学出版社
http://www.hustp.com
中国·武汉

目录
Contents

现代酷宅
Modern Cool Mansion

美式阔宅
American-style Mansion

南非华宅
South African-style Mansion

现代酷宅

Modern Cool Mansion

Artistic Life: A Profuse Presentation

艺术生活，缤纷呈现

项目名称：私人府邸 1
设计公司：ODA 建筑事务所
摄影师：弗兰克·乌德曼
地点：纽约
面积：167 m²

Project Name: Private Residence I
Design Company: ODA–Architecture P.C.
Photographer: Frank Oudeman
Location: New York
Area: 167 m²

"私人府邸"为曼哈顿最大的公寓楼之一。90层楼的高天之天，全方位的开放，360度的视角，静观纽约城市的天际线。诺大的空间也成了世界最大的私人艺术博物馆之一。宏伟的空间，以主人的"纽约漫行"为主题。

步入其间，首先经过的是一个"禅意"花园。动线设于东部，风景独好。东河、皇后区、布兰克林区、机场、联合国大厦，一一映入眼帘。居家的生活如同漫步于纽约。

花园位于东部，恰好凌驾于东河之上，不但连接起东河与空间中央的泳池，还借助于瀑布的设计，引水直上。借助于主轴，空间一分为二。南面为隐私区，其余为娱乐区。娱乐区设有影音室、品酒室、雪茄室、投影室、休息室等等。

属性完全不同的用材、质感运用于整个空间。徜徉其中，感受到的却是一个与众不同的业主。

This particular apartment is one of the largest apartments in Manhattan. It's totally open on the 90th floor with 360 degrees views of New York City, housing one of the largest private art collections in the world. The central theme of the space was the journey of this one family to the city of NY, this epic place.

As you enter the apartment, you cross what we call the Zen garden. We chose to enter the apartment facing the East River, Queens, Brooklyn, the airports, the U.N. Building, as a symbol of the gateway, the path this particular family traveled, which allowed us to really create a rich and deep sequence of cultural journeys.

The idea of placing the Zen garden on the east side, right above the east river, thereby connecting visually the horizon of the river with the reflecting pool at the center, and then bringing it vertically, through a waterfall, to the upper floor, is the kind of thread that we have tried to create throughout the apartment. What we decided to do was literally divide the apartment into two parts that connected through a main spine. The southern part of the apartment is the private residence. The rest is the entertainment part, which includes the listening room, wine room, cigar room, the lounges and a projection room.

Totally different materials and textures are employed throughout the apartment, and as one goes from one room to another, every room expresses something totally different about the client.

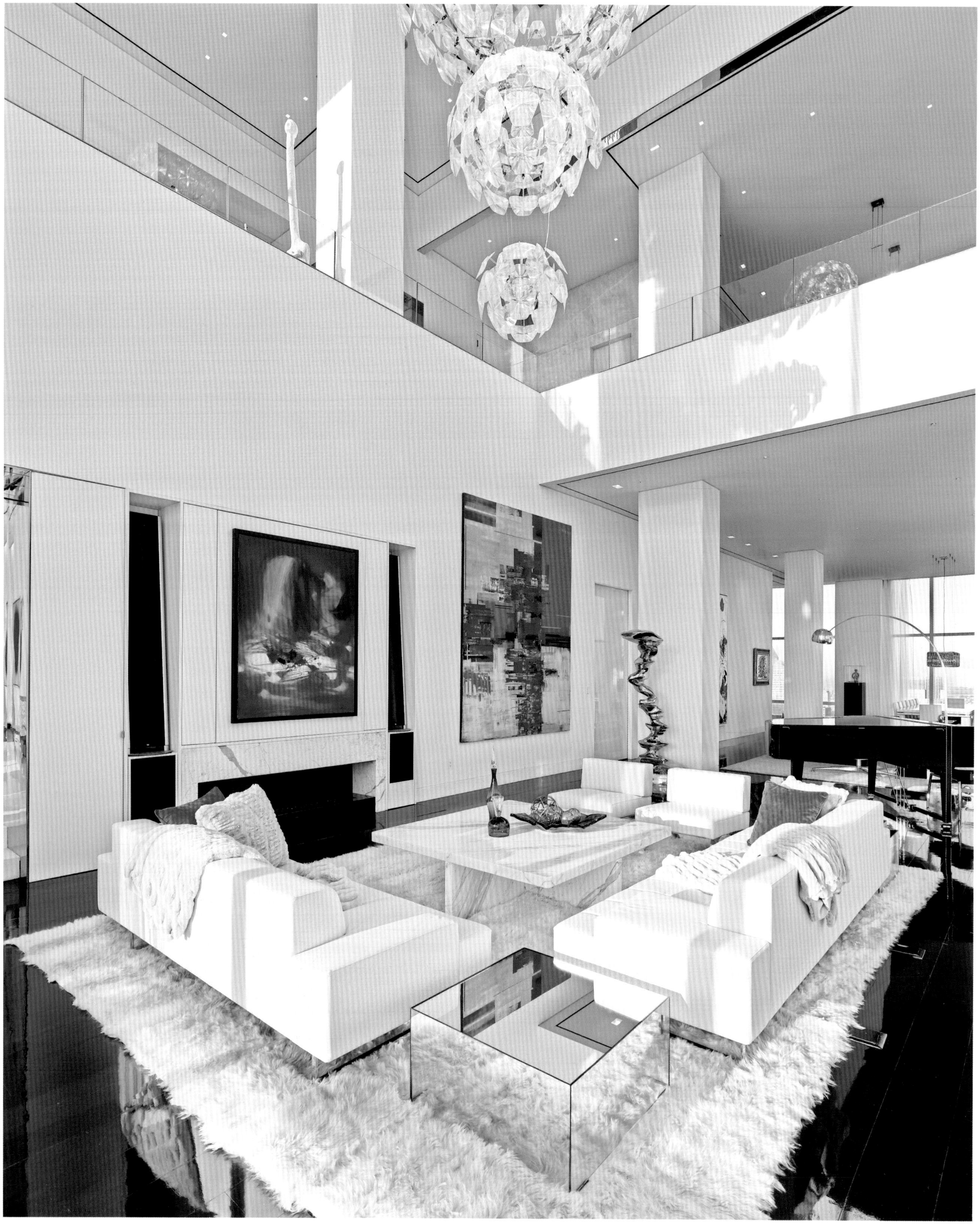

A White House with Hills in Front and Sea at Back

依山面海，充满绮丽想象的白色童话屋

项目名称：洛杉矶"凡尔赛"宫殿
建筑公司：比格林建筑
建筑师：布莱恩·比格林
面积：3 255 m²

Project Name: Beaux Art Beverly Hills
Construction Company: Biglin Architectural Group
Architect: Brian Biglin
Area: 3,255 m²

"白鸟别墅"，中轴式设计。其轴线便是动线。沿着动线前行，入鸟尾部，至头部，出了鸟嘴，便是大海。睡眠区，隐密于中部，其他各处分设着公共区、厨房等各服务空间。

卫浴的设计，以其所提供的感官享受吸引着你步入其中。移步换景式的空间，给人别致的体验。接待区有一旋转楼梯。沿着楼梯，可进入个性斐然的各个空间。每一个空间环形设计，自成世界。对称着立面，环形的地板与屋面之间悄悄地进行着对话。直线式的屋面下，是一个家居的天地。无阳光的暴晒，无风雨的干扰，家居尽在和煦之中。

"鸟"的腹部，是生活起居室。头部、心脏、嘴部都各有设施。翼部，设有双阳台，可以用作生活起居室。各空间无不依地形而建。花园位于前面右部，泳池位于中部。水声潺潺给人以温暖的享受。寒冷的冬季，泳池适时转变成灯具，灯光投射于天空，映射出鸟的映像。园艺大师布莱恩（Brian Wesley）按照本案空间的需要，以其精细的视角，专为本案定制了一个中央草坪。沿草坪排列的各处空间，错落有致，给人一种天堂的观感，如梦如幻。

Being a house with a defined axis, which is the access, we can say that we come in through the bird's tail, and out through its head, towards the sea. To the sides there are the sleeping areas, private in the center and the common areas, kitchen and service areas are distributed.

You walk into the bathroom and guess the sensual and suggestive forms that invite you to access. That attraction is discovered in stages, without a complete picture. First, a reception area with a spiral staircase that lead us from place to place quite naturally into rooms designed for a family with many children. Each one with its own world, its personality and organized around a circular. Against the facade there is a dialogue between the circular floor and the roofs that go in a straight line. A roof that protects. A roof is a home. It protects from the sun, rain, gives warmth

In the guts of the bird, there are the living rooms, and the heart, head, beak, come in a special harmony as extras. In a wing, the terraces double the living room. But the distribution is determined by the topography. The gardens are also at the front right and pool, symbol of the holiday, in the central area, very close, very immediate, very present. It offers warmness, freshness, sound. In winter it is a great source of light, the pool turns into a lamp, into an inverted sky, where the birds is reflected. Gardener Brian Wesley was the artist able to understand the needs of the place. He has very subtle vision of the play. He has created a large central lawn area, staggered spaces at the sides converted into paradise. A dream that gives a very good impression.

Movement by Art Goes Gorgeous Everywhere

艺术谱就的华丽乐章，响彻生活每个角落

项目名称：以色列别墅
设计公司：OE 建筑师事务所

Project Name: House in Herzelia, Israel
Design Company: Oded & Elizabeth Tal Architects

本案基地狭窄，17 米宽，但纵深达到46 米。因为泳池位于空间尽头，设计因此强调线性设计。脚步一路向里，便可直达花园。深灰色的火山岩外表，横向镶嵌着窄窄的条纹。整个量体分为前后两部分，中间有一个三层中庭相连。中庭的四周用玻璃作为围护，北部便于采光。中庭的焦点为一圆形楼梯，并有步道，连接着前后空间及楼上的卧室。

除了一楼的四个卧室套房外，拱形的屋顶下还有一个主卧。主卧配有阳台及冲浪浴缸。地下室设有家庭影院及三个客房。

空间以黑、白二色为基调。黑色的铝合金框架对比着白色的铝制屋顶。室内地板由白色天然石切割而成，长宽约为一米。浴室使用的是本地灰色石头，并饰有各种纹理。定制的家具、华美的色彩对比着中性白色的墙壁。

We were approached by the owners to design a house on a long, narrow site which was 17 meters wide along the street and 46 meters deep. With a swimming pool located at the far rear of the house, the design principle was to emphasize the linearity of the house and draw the visitor all the way into the garden. Hence the exterior finish of dark grey volcanic stone in narrow horizontal strips. The volume of the house is split into front and rear wings connected by a three storey atrium enclosed by curved glass walls, allowing volumes of natural light from the north side. The focal point of the atrium is a round staircase and bridges which pass through the void connecting the front and rear bedroom wings on the upper floors.

In addition to the four on-suite bedrooms on the first floor, within the vaulted roof of the attic is a master bedroom suite adjoined by a sun balcony with whirlpool tub. The basement houses a cinema room and three additional guest bedrooms.

The house is finished in a palette of black and white. Black aluminum frames the windows in contrast to the white aluminum roof. The interior flooring is white stone cut in 1.0 x 1.0 meter blocks and the bathrooms use various textures of local grey stones. The furnishings were custom designed and selected to be very colorful against the neutral white walls.

沉醉于当代艺术的无限创想

Simmered in Creative Infinity by Modern Art

项目名称：NS 别墅
设计公司：加莱亚佐设计

Project Name: NS House
Design Company: Galeazzo Design

本案位于巴西圣保罗，600平方米的空间，以艺术、设计为家人提供一个家居、会客的场所。空间采用自然、可持续的材料。地板所用木材、石材，为其他建筑拆卸回收利用。社交区分为生活区、酒窖、美食厨房、阳台等几个区域。陈设除了古董家具还有意大利、法国等品牌产品。藏式地毯铺于整个空间。

阳光透过大大的门窗，盈满室内。扇形的布局结构，以开阔的动线相互关联。即便是客人，在融入空间的同时，隐私也可以得到充分的尊重。美食厨房气场恢宏，内部各功能区间相互融通。夜晚的派对，这里是当仁不让的主角。

深深的庭院，幽幽水景，升华着、放大着城市绿洲的感觉。

Located in the city of Sao Paulo, this 600 square meters house was designed to a family with the aim of being a place of social meeting among a collection of art and design. For coating it was chosen naturals and sustainable materials like demolition wood for the floors and raw marble. The social area was divided in sectors: Living, Cellar, Dining room, Gourmet Kitchen and Balcony. For decoration, vintage furniture mixed with Italian, French and Scandinavian design furniture. In all rooms there are Tibetans rugs made of wool and silk.

The house was conceived with big windows and doors that allows daylight to be the big supporting element of the spaces. Despite the sectored ambient, all the house are integrated through spacious passages allowing the guests to enjoy the spaces with privacy but integrated to the house. Spacious passages allow the guests to enjoy the spaces with privacy but are integrated to the house. The gourmet kitchen is the big venue where all spaces are joined together and it becomes the main attraction on the party nights.

In the big yard there's a pool for the summer days that magnifies the effect of oasis in an urban setting.

Mansion of Concrete to Accommodate Seasons

混凝泥质朴大宅，收藏生活的四季风景

项目名称：五子别苑
设计公司：马丁·戈麦斯建筑师事务所
设计师：费德里科·罗德里格斯、多明戈达·米兰达、佛罗伦西亚·罗德里格斯、佛罗伦西亚·里帕
摄影师：埃塞基耶尔·伊斯克兰特
地点：乌拉圭

Project Name: La Boyita
Design Company: Estudio Martin Gomez Architects
Designer: Federico Rodriguez, Domingo Miranda, Florencia Rodriguez, Florencia Rippa
Photographer: Ezequiel Escalante
Location: Uruguay

本案空间由五个量体构成。主量体设以起居和餐厅空间。建筑面朝大海、泳池，自是尽览荡漾水波。伴着微风，空气中皆是水的清凉。其他的量体设有服务区、客人房。客人入内，无丝毫外界之干扰，尽享宾至如归之感。

各量体周围，绕以回廊。室内生活空间至此延伸至外。全景式的内外空间顷刻间没有内外之间的界定。空间用材，种类繁多，诸如混凝土、木、铁、玻璃。

The project consists of five blocks which contains a main block with a living and dining area with the best views to the sea and its swimming pool, which is completely protected from the wind placed in the center of the project. The other blocks contain a service area and other guest room blocks. This independence was crucial to make his guests get the feeling of being in alone in a hotel.
The corridors that surround these concrete blocks end up creating the idea of outdoor living rooms and creating a fluid transition between the interior and the exterior with a fantastic panoramic view. The materials used are concrete, wood, iron and glass.

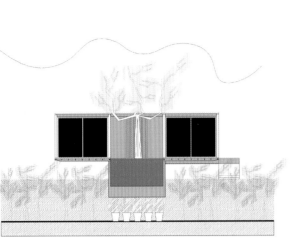

立面图 / Elevation Drawing

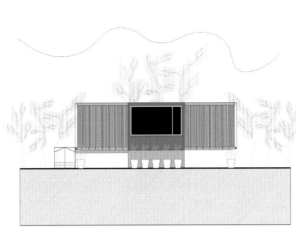

立面图 / Elevation Drawing

平面图 / Site Plan

Beijing Fantasy

京城幻想曲

设计公司: Dariel Studio
设计师: Thomas Dariel
面积: 1500 m²

Design Company: Dariel Studio
Designer: Thomas Dariel
Area: 1500 m²

Dariel Studio 最新完成的私人住宅项目可谓是奏响了一部现代感的幻想曲。这个 1 500 平方米大的公寓坐落于繁华的北京三里屯地区，超凡的装饰设计完全体现出业主不凡的性格。

楼顶最上面的两层总共 12 个小公寓被全部打通组合成一个复式公寓，通过创造大容积的布局，设计师 Thomas Dariel 给予了这个室内设计巨大的空间感。

制造出开放性的空间是设计首要解决的。一楼就是一个巨大的开放区域，没有任何隔断，也没有保留墙体，而是运用不同纹理、材质、颜色、线型和造型来区分不同的空间，让每个空间诉说不同的故事。

超大挑空的客厅空间，由纺锤形的承重柱支撑二楼的结构，可以反射一二楼的镜面包裹的横梁，让人很难知道空间的连接处。由于入口处的天花太低，设计师也运用了视错手法带来了同样的空间感。走入时，深色的木地板将客人引进主客厅，而周围螺旋式黑白条纹的互相反射，制造出迷幻的氛围让人无法分辨处于哪里。透过入口，可见似乎是无止尽旋转的圆形楼梯，令人印象深刻。这个旋转楼梯本身就如同艺术品一般，位于整个空间的中心，以开放式的姿态连接着每个功能区域。它是整个空间结构的精髓，也是整个设计的心脏。

Thomas Dariel 的空间布局方式也奠定了整个设计的基调。这个新项目向后现代主义致以崇高的敬意，其明亮强烈的色彩、装饰性的表面纹饰、不对称的线型和形状都营造了一种奇特而有趣的氛围，这些室内设计的元素都具有后现代孟菲斯运动的特质，其中更是受到 Ettore Sottsass 的灵感启发。"我被孟菲斯运动和当时的设计影响了，我将这些在我儿时记忆中产生的积极影响运用到了我的设计中"。

整个空间的设计是优雅而精致的，所有天花、墙壁、柜子、家具都是私人定制的。Thomas Dariel 将空间打造成一个既能带给人真实体验又能使人感到舒适的家。各种颜色碰撞出的火花实现了这一设计初衷。对设计师来说，色彩是表达情绪的最好媒介，通过对应的色彩可以很好地定义不同的功能区域。

"眼见的外表形式只是带来愉悦，而真实的本质才带来生活"（蒙特里安）。源于生活，Thomas Dariel 在此给出了一个有趣味的、奇特而诗意的答案，将生活的烟火气藏进了一个童话世界。

风格化的当代艺术作品不时地点缀着整个空间，带来了无与伦比的艺术气息，也更好地反映了客户热爱艺术和设计的生活态度。入口处摆放的由艺术家 Aurele 带来的作品"Lost Dogs"，与黑白条纹所营造出来的当代感相得益彰，使人仿如置身某个艺术展的错觉。而其他的区域放置的艺术作品又带来完全不同的气场，如一组知名设计师 Claudio Colucci 的"Squeezes"灯光作品，气流线型的造型柔化了大空间的空旷感，营造了独特而不张扬的平衡之美。

如果说设计风格是如此的夸张，但公寓的布局却完全是基于客户的需求之上。一楼更多的是满足公共需求，如玄关、客厅、餐厅、厨房、客卧、客卫、儿童玩乐区和艺术陈列区；二楼则是更为私密的房间，主卧、主卫、儿童房、浴室、家庭区、更衣室和书房。每间房间布局和整体都由设计师协调设计以保证一个和谐舒适的住宅环境。

"这个项目像是我 30 岁之前的设计的最终篇章，它集合了我最标志性的设计风格和语言，也很好地表达了我是怎么样的设计师。"

The last on-to-date residential project designed by Dariel Studio is a truly modern fantasy. Located in the vibrant Sanlitun area, this 1 500 square meters apartment features an extra ordinary decor that pays homage to the client's eccentric personality.

12 apartments spread over the two last floors of the building have been reunited to give shape to the current penthouse. Thomas Dariel 's new interior creates volumes that offer a sense of space, a feeling of immensity.

Openness is indeed the first statement. The first floor is a vast open space. No partition. Instead of walls, Thomas Dariel plays with textures, materials, colours, lines and shapes to create spaces that tell each a different story.

The living room features a huge void with supporting spindle-like columns and transversal mirror-covered beams that reflect first and second floors at the same time. Looking at the reflection, ones have difficulty to know where the space starts and ends. While having a rather low ceiling, the entrance also uses an optical illusion that brings the same spatial feeling. When entering, a dark wooden path guides the visitor through the main hall, surrounded by spiralling black and white glossy reflective stripes. Ones

have difficulty to position themselves in this psychedelic space. In the perspective, an impressive and once again seemingly endless round-shaped staircase yet helps the visitor to find his path up to the core of the apartment. Masterpiece in itself, the staircase is located at the center of the penthouse, opening, connecting and leading to all rooms. It is the pith of the space structure, the heart of the design.

The way Thomas Dariel plays with the space sets the tone for the whole design. Indeed, this new project is a clear homage to the extravagance of post-modernism. Bright and strong colors, decorative surface patterns, asymmetric lines, shapes that have little reference to function, deliberate eccentricity and playfulness, all these elements featured in the apartment's interior design are very much characteristics of the post-modern movement known as Memphis and represented, amongst others, by Ettore Sottsass. "I was influenced a lot by the Memphis movement and the designs from that period, these are positive memories from my childhood which I wanted to bring them back into the design".

The whole space is elaborated and sophisticated, featuring an utterly designed environment where ceilings, walls, storages, furniture have been customized to serve this design purpose. Thomas Dariel creates spaces that offer a real experience and that make people feel at ease. Colors are strong tools to

achieve that goal. Always associated with emotions and feelings, colors help to define the space functions and their related sensibility.

"The surface of things gives enjoyment, their interiority gives life". (Mondrian) Life. Thomas Dariel designs here a playful, eccentric and poetic décor that tells us a story, bringing life to a fairy tale. Contemporary art pieces are displayed throughout the space, reflecting the client's passion towards art and design. The "Lost dogs" from Aurele, placed right at the entrance as if ones were entering an art exhibition, complement and contrast with the firmly modern atmosphere. In the art display zone, the emblematic Claudio Colucci "Squeezes" are subtlety enlightening the space with gracefulness and mystery.

If the style is extravagant, the master plan of the apartment answers the client's needs and requirements. While the first floor resumes the public functions of the apartment – entrance, living room, dining room, kitchen, guest bedrooms, guest bathrooms, kids playground and art display area - the second floor accommodates more private and intimate ones – master bedroom and bathroom, kids bedrooms and bathrooms, family area, dressing room and study. Each space itself and the way they related to one another were thought and designed in collaboration with a Fengshui master to ensure smooth flow and harmony.

"This project is for me then ending chapter of my design before my 3os. I gathered all my design signatures, references and styles and expressed them through a design, referring to who I am."

Color of Life in Era of Science and Technology

科技时代的彩色生活

项目名称："白宫"
项目位置：法国拉罗歇尔
设计公司：皮埃尔公司
摄影师：亚瑟·帕昆

Project Name: White House
Location: La Rochelle, France
Design Company: Pierre-Antoine Compain
Photographer: Arthur Péquin

"白宫"设计围绕全高中庭展开。白天的日光透过巨面的天窗，明亮地照耀着四层空间。玻璃透明的隔断，白色的地板，让空间无论横向、纵向都有一种通透感。各功能空间，如电脑房、阅览室等，如同"豆荚"一般，小而精致。

白色的基调，伴随着光照，让空间的色彩有了一种"爆炸"的感觉。玻璃铸就的"轴承"从上到下，连接着包括中间平台在内的上下空间。玻璃"轴承"在美化空间的同时，更起到 "温室"的作用，佑护着来自异域的植物。开放的屋顶是空间的采光井，让整个空间的通透感和光感极佳。

The scheme is organized around a full-height atrium, located beneath a large roof window, which allows light to flow into each of the four floors. Glass partitions and white floors in this inner void promote illumination, while promoting an overall openness both in vertical and horizontal axes. Small pods extend into the central space to create areas scaled for individual use, including a computer station and reading nook.

The dial tones of white married with the light, make it possible the color to explode. The main element of White House is very clearly the glass axis which crosses it from the upwards. In reality, this axis is a greenhouse, piercing the ground floor to join an intermediate landing, before the first floor, the greenhouse receives exotic plants. All the house lives thanks to a game of transparency and light, strongly strengthened by the opening of the roof which creates a light shaft essential to the balance of the project.

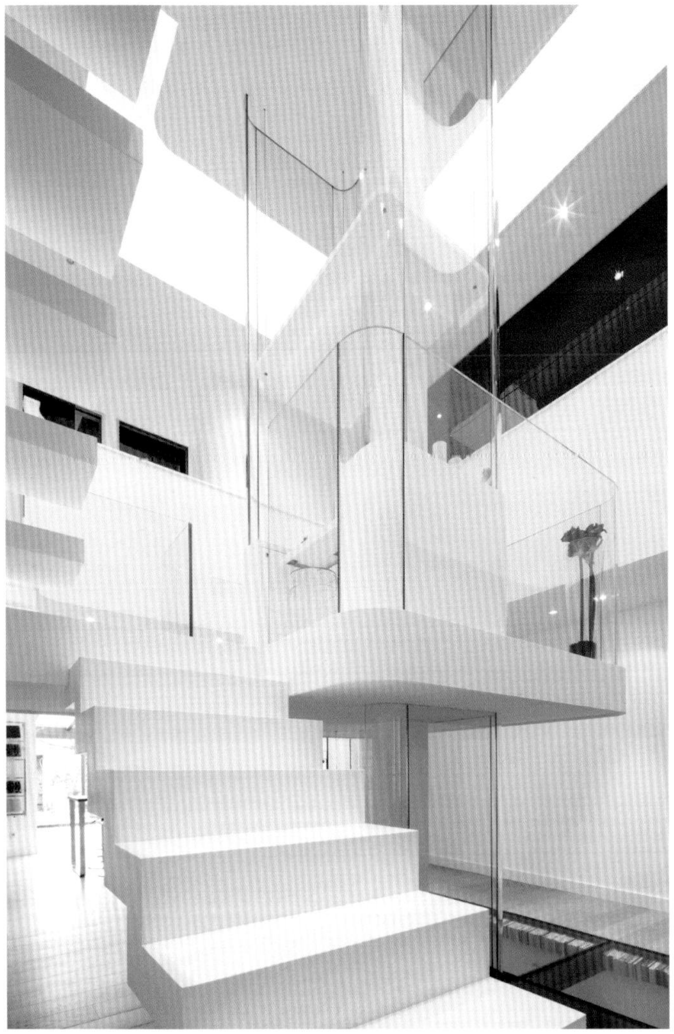

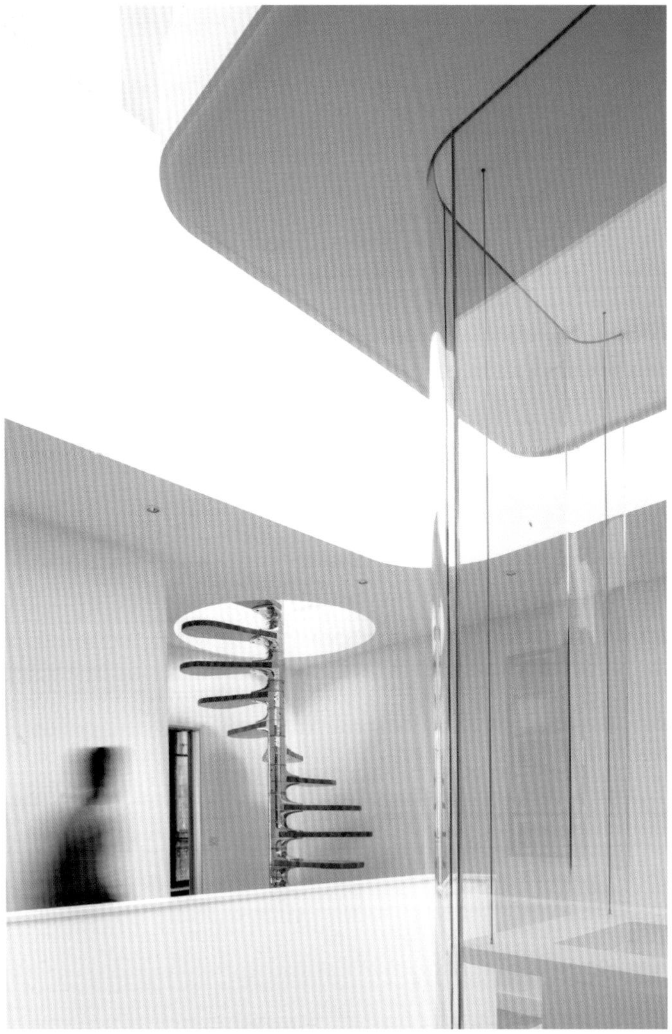

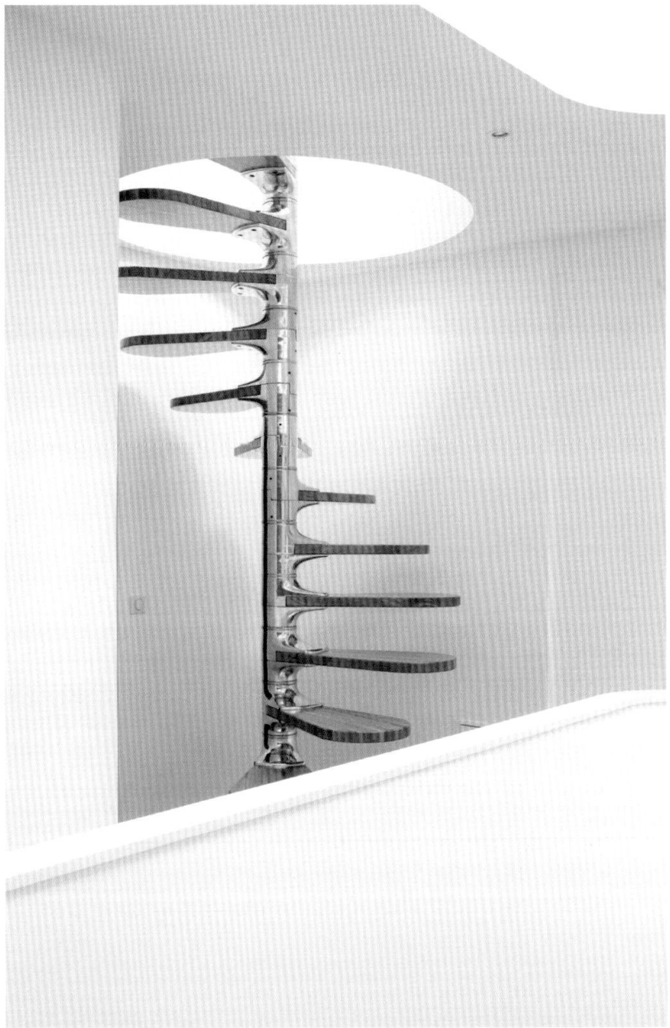

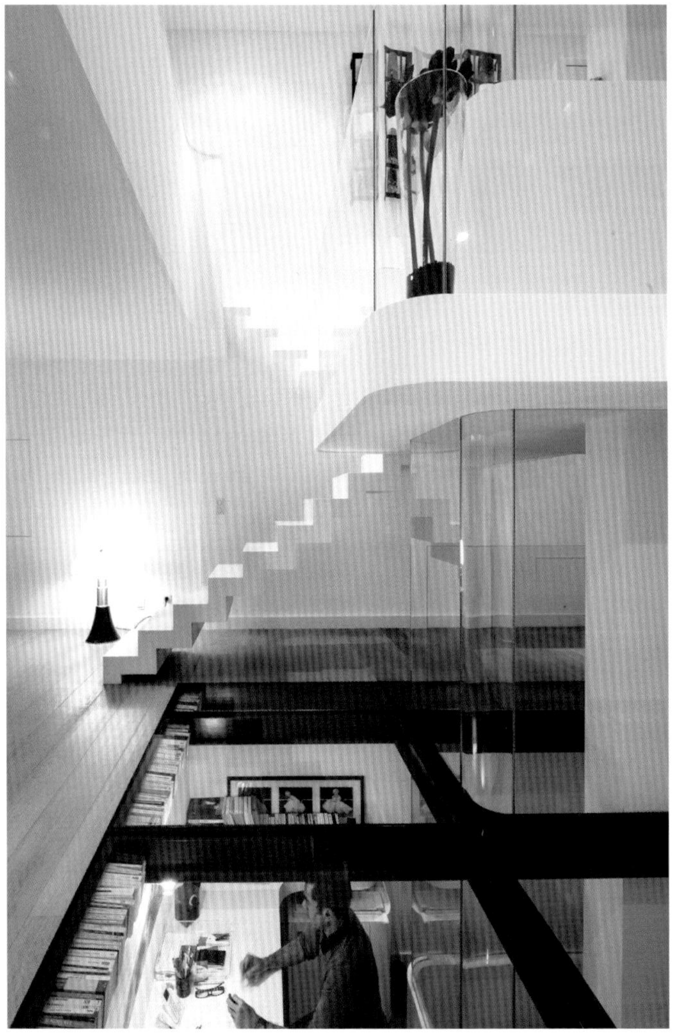

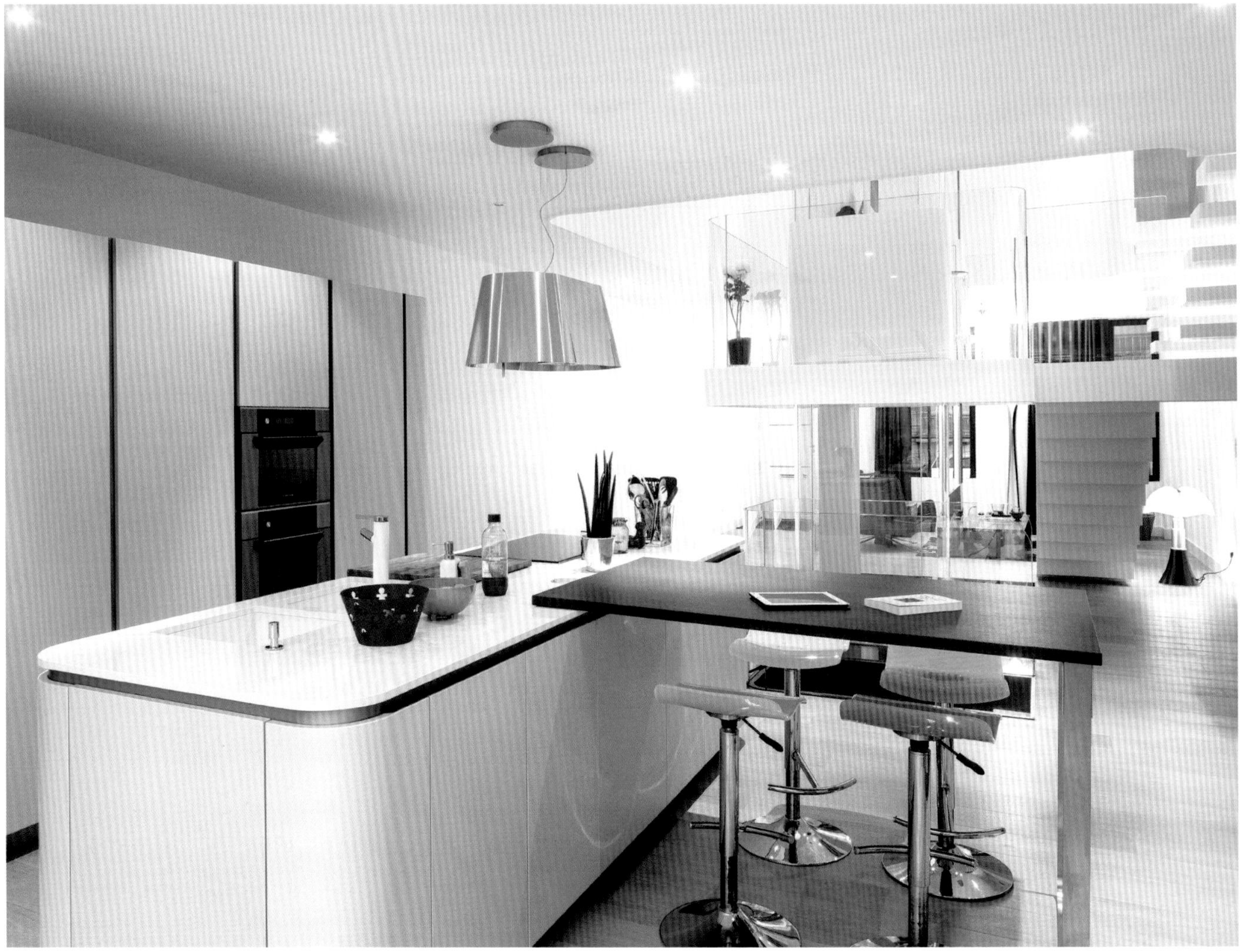

Dwelling Fine and Refine on Hillside

依山就势而建，纯净雅致之居

项目名称：加州别墅
设计公司：麦克林设计
地点：加利福尼亚洛杉矶

Project Name: Sarbonne Rd Residence
Design Company: Mcclean Design
Location: Los Angeles, CA

住宅位于洛杉矶，占地1 027平方米，基地有些陡峭，设计颇有难度。设计的目标就是希望利用可用的土地，创造一个温暖、现代的家居生活空间。健身泳池虽小，但于本案也做到了极致化。泳池与建筑之间的平地形成了一个很好的缓冲。房子下面设有一个大大的车库。楼梯、水墙直通入口，形成了一个环形的通道。一楼除了几个娱乐室外，还有卧室、图书室、多媒体房与酒水间。

This is a large home of approximately 1,207 square meters located on a steep and difficult lot in Bel Air, Los Angeles. The goal of the project was to create a large warm contemporary home and to maximize the flat area of available land. The pad is extended by pushing the lap pool to the extremities of the lot and creating a flat area between the house and the pool. A large garage is located below the house creating a circular entry court with a stair and water wall leading the way to the entry. The house consists of several entertainment rooms on the lower level and incorporates several bedrooms as well as a library, media room, and wine room.

Fairy-Tale House of Five Colors in Forest

森林里的五彩童话屋

项目名称：轩
设计公司：SA 建工
设计师：彼特·斯坦伯格等
摄影师：保罗
面积：102 m²

Project Name: Shelter Island Pavilion
Design Company: Stamberg Aferiat Architecture
Designer: Peter Stamberg, Paul Aferiat, Keith Tsang, Joshua Homer, Ryan Harvey, Josh Lekwa,
Anna Portoghese, Michael Bardin, Adam Greene, Jasmit Rangr
Photographer: Paul Warchol
Area: 102 m²

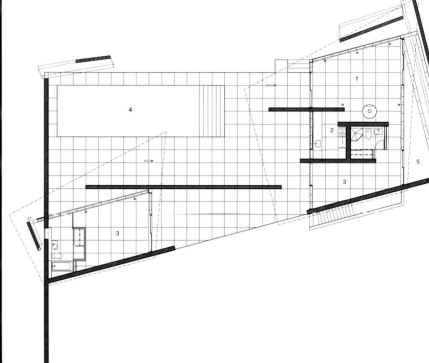

平面图 / Site Plan

"轩"空间是一种开拓，以工业化的用材及与时代同行的方法实现了一种大胆尝试。不再使用特殊的用材，空间尽以普通的材料，但其用法大胆、创新，色调精心搭配，细部考究。

立体主义的视角已经超越了机械的观点。视觉映像与设计匠心相互辉映，众多映像相互之间看似有些脱节，但却具有整体的含义。轻型建材的使用，让建筑形式直截了当地呈现出来，也让诸多的建筑梦想成了现实。盘旋的顶，通透的空间，似是而非的墙，让概念就这样变成了现实。

现代的材料技术增加了色系的浓度，提升了光学的效果，升华了各种运用，空间的调色板因此有了扩大。有了牛顿色彩理论的武装，透明墙体间多了更多的光影舞动，空间到处弥漫着一种兴高采烈的光。

除了严格的调研及科学理论作为武装，整个设计极其讲究用材、方法的可持续性。所有功能空间一年四季皆可利用。制冷、取暖可应时而变。推拉门的开启闭合成就着夏冬两季室内空间的开放与封闭。本案空间是当地第一批使用地热资源进行供暖制冷的建筑。有了精妙的设计，即便炎热的季节，制冷似已成了一种摆设。西、南两面的实体墙阻隔着太阳的炙热。东、北两向，全高的透光树脂墙和传统的玻璃墙相比，热阻性能更好。西、南两面还配有推拉门、阔大的窗，以便通风、采光。宽大的挑檐为内部提供了阴凉，也提供了理想的户外生活空间。

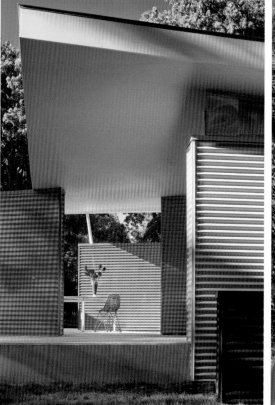

Stamberg Aferiat designed a house that explores the reality of the industrially-produced materials and methods of our own time. Instead of then-exotic materials, they chose to utilize more common materials but rendered them striking in usage, pigment choice and detailing.

Cubists looked beyond the mechanical view of how the eye sees and employed the brain's ability to remember and anticipate, allowing one to take in a seemingly disjointed array of phenomena but still have the whole make sense. The increasing plasticity of lightweight building materials allowed Stamberg Aferiat some of the Cubists' slight-of-hand to simultaneously evoke the immediacies of built form as well as architectural dream states – the hovering roof, translucencies between inside and outside, and walls that are not walls.

Advancing material technologies have expanded the available palette through increased color intensity, optical effects and applications. Guided by Newtonian color theory, the intense palette of the house allows richly-colored reflected light to pass through translucent walls, suffusing spaces with a delighting glow.

In addition to the rigorous studies of perspective and color theory, environmentally sustainable materials and methods played a large factor in generating the design. The home is designed for all seasons with the use of the spaces and the areas conditioned are modulated based on seasonal weather. Its heaviest use is during the summer. Large sliding doors allow indoor functions to flow into outdoor terraces and gardens during the summer when additional space is desired and indoor conditioned space is rarely needed. The opposite occurs in winter where living occurs in a much smaller conditioned footprint. The house is one of the first on Shelter Island to use geothermal heating and cooling. Even so, it is rarely used in the warm season as Stamberg Aferiat incorporated many passive design elements into the architecture. Solid walls on the south and west side of the building block the intense summer sun while floor to ceiling translucent double polycarbonate walls allow north and east light into the space as well as providing a much higher R-value as compared to traditional glass. Large sliding doors and windows are carefully placed to take advantage of the east/west sea breeze to cool the interiors. Large roof overhangs provide needed shade for the pavilion interiors while providing sheltered space ideal for outdoor living.

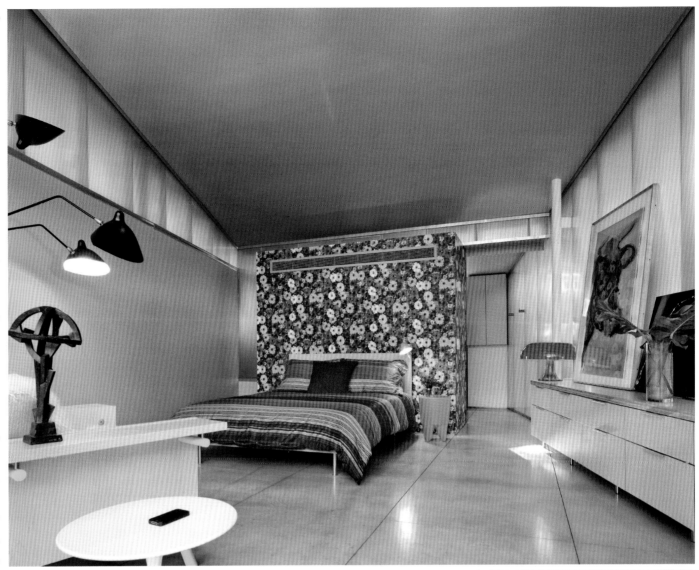

定制优雅，现代科技成就时尚生活

**With Custom Elegance,
Modern Science Makes Life Fashion**

项目名称：鹿特丹公馆
设计师：罗伯特

*Project Name: Rotterdam Residence
Designer: Robert Kolenik*

在业主的全权授权下，鹿特丹公馆不仅时髦，而且舒适。

客厅里的大沙发，长6.5米，极为惹眼。细打量，原来是本案设计的定制作品。透过旋转的门，客厅成了一个艺术品的世界。所有艺术品，是本案设计的收藏。厨房的桌面由纯白色玛瑙制作。桌面的支撑是荔枝树的树干。

卧室也是不同凡响，自成一个奢华的套间。卧室里有步入式衣橱，衣橱的把手蒙着亮光的鱼皮。卫生间给人一种健康、阳光的感觉，里面有两个完全一样的黑色玛瑙水槽。

厨房的角落立着的地灯出自本案设计师之手，其他的空间照明由别人专业制作。地下室里有酒吧、影院。本案设计自行制作的音响系统给人从视觉到听觉的全方位享受。所有的电器，全是自动化设施。水泵是最新的流行技术。而雨水的收集利用使整个空间变得更加有持续性。

多功能的书桌，长宽皆为2.5米。因为多媒体的出现，坐在桌前办公，便成了一种享受。有墙体以"皱纹"的形式出现，同时镶嵌着黑黄檀木的组柜。在这里，除了快乐，更有享受。

What do you get when you are the owner of a sublime villa in Rotterdam and you give designer Robert Kolenik carte blanche to redesign your interior? Eco chic eye candy for extremely comfortable living.

Take the living room for example. With that grand, eye-catching, custom-made Proud couch by Kolenik measuring 6.5 meters in length, together with a pivoting door from Kolenik's own Bo'dor collection. Sheer happiness is expressed by the kitchen with its illuminated tabletop made from pure white onyx and supported by a lychee tree trunk.

Kolenik's global boutique design studio has pulled out all the stops with the master bedroom as well; an en suite walk-in-closet with flush fish leather handles and a divine bathroom that embodies the wellness experience and boasts twin lit sinks in black onyx.

Illuminating the villa is Maretti lighting while Kolenik's own floor lamp 'Dream' radiates out from the corner of the gorgeous kitchen. In the basement you'll delight in a bar and cinema, with a Kharma sound system to make the entire house tantalising for the ears as well as the eyes. All the electronics in the villa are automated by Bits & Bytes, while the latest techniques in water pump devices and the use of

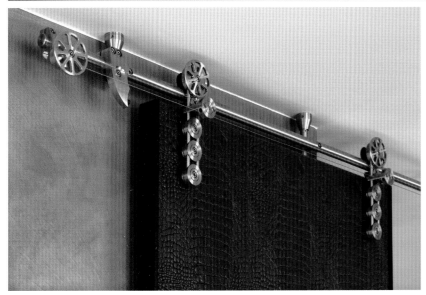

rainwater make the house more sustainable. Work becomes a pleasure when sitting at the 2.5 x 2.5 meters multifunctional desk with integrated multimedia, particularly with one wall clad in "Wrinkle" by Dofine and adorned with palisander wood cupboards. A perfect example of what you get when you give Kolenik carte blanche: happy clients.

Spatial Expression Completed by Light

光赋予空间生动表情

项目名称：台北信义 A-3F
设计公司：界阳 & 大司
设计师：马健凯
面积：270 m²
用材：不锈钢镀钛、LED 灯、钢琴烤漆、石材

Project Name: A-3F of Faith, Taipei
Design Company: Jie Yang Interior Design
Desigher: Ma Jiankai
Area: 270 m²
Materials: Titanize Stainless Steel, LED Lighting, Piano Stoving Varnish, Marble

本案以罗浮宫拿破仑广场上的玻璃金字塔为设计灵感，将设计理念融于一点、一线、一面之中，以连续性的美学风格，超越对住宅的认知，加深艺术雕塑感，让前卫的辨识度相形而生。

层序玻璃光束引领入室，通过绚丽的星光大道与融合以简驭繁的当代观点的白色金字塔，时尚的线条与异材质的交汇，形成不规则的空间边界，让几何语汇串联至立体切面，首尾同时呼应雕塑主轴，更见线条挥洒自若的流畅张力。室内灯光氛围依旧是一大亮点，魔法般切换出艺文博物馆、时尚伸展台、Lounge夜店等情境，让艺术和精品生活奢侈并存。

本案拥有得天独厚的62.7平方米的超大露台，自成尘嚣不扰的城市角落。发光体至餐桌、吧台，融合积木堆栈概念，串接出客厅、餐厅、厨房的另一种形态。考虑家具置身露天环境，特从国外进口百年不朽的特殊地板，以及历经风吹日晒亦不损坏的台面质材，构筑空中花园与Lounge Bar的时尚意象。发光的柜体与镶嵌光沟的台面，以灯光改变露台情境，烘托出时尚前卫的夜店氛围。

走入私人领域恍若穿越了时空之门，眩晕的光感延伸出无尽的时空隧道，奇幻层次止于末端的多功能平台，钢琴、爵士鼓、单杠等休闲设备一应俱全。数间卧房，以空间本质为主导，洗练、时尚或轻奢，都让艺术品位不着痕迹地蕴藏其中。

夜幕低垂时分，阳光树梢打破根深蒂固的思考逻辑，融会艺术理念与光影层次，让价值与品位诉诸五官，随光、影、时序而变，空间情境自然生成。

在书桌区，展示架使空间有区域之别，同时纳入树梢绿景与自然采光。书桌以3D光束屏风为支撑，轻与重的结构产生冲突美感，更显前卫。收藏城市夜景的角窗内，以书桌延伸、串联洗脸面盆，并于转折处搭配水滴形灯具，凝聚角落的精彩。

主卧房外拥树梢绿意、101大楼景致，床头以皮革搭配

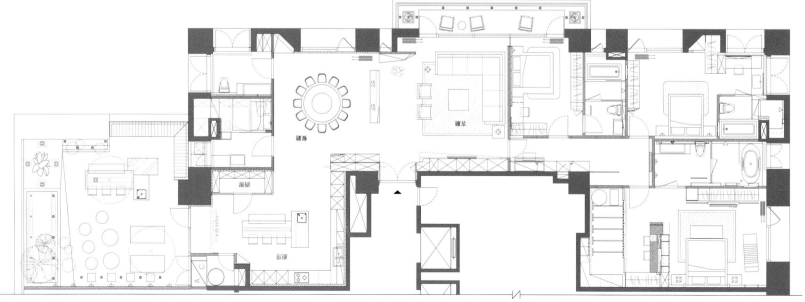

平面图 / Site Plan

SWAROVSKI 水钻，黑白色调匹配出时尚轻奢华。

Inspired by the Glass Pyramid on Napoleon Square of the Louvre Museum, artistic concepts are reclusive in spots, lines and surfaces to present a continuous aesthetics for further enhancing sculptural idea. This gives birth to a leading recognition.

The sequence of glass light beam leads into the interior. Opposite to the wall, the aisle with its form of pyramid express a contemporary approach that the simple is able to control the complex. Lines spreads in communication from the flooring, to the wall and to the ceiling, making an irregular boundary and completing 3D sections of all kinds. The head-tail correspondence brings forward a spatial sculpture effect while bringing out a flowing but composed texture. The interior lighting makes another spot light, completing images of artistic museum, extending table, lounge and night Entertainment Avenue, so art and boutique life can coexist.

It's large terrace is incomparable, offering a corner far away from the hustling and bustling in a city setting. Shiners of dining table and bar are employed with building block concept to string the living room, the dining room and the kitchen. The flooring is particularly exported, whose material quality has been immortal for dozens of years. The table-board quality can be endurable after being exposed to the weather. All makes up images of air garden, lounge and bar. The shiny cabinet and the inlaid-lighting table surface set off the

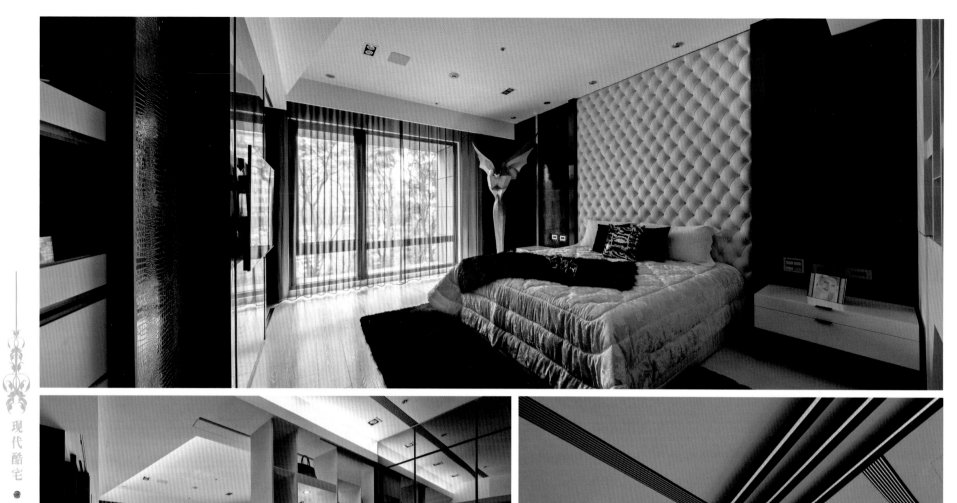

atmosphere of the night entertainment avenue with light timely to change the situation.

The entry into the private room seems to have been through the time tunnel. The bitingly cold light sensation extends into numerous spatio-temporal channel. The queer comes to its end when the multi-purpose platform, piano, drum set, and horizontal bar come into sight. Bedrooms, dominated with spatial nature, are either sophisticated, fashion or somewhat luxurious. All are of art taste but low-key and conservative.

Whether sunlight above treetop, or the night setting, the presentation breaks away the inveteracy thinking logic and fuses the artistic concept and light layers, so that value and taste resort to five organ senses and change with time and light and shadow varying. Spatial situation is simultaneously generated.

Around the reading area, the bookshelf serves as partition taking in landscape and daylight. The 3D light-beam screen supports the desk, making a contrasting aesthetics in the light-heavy structure in setting off another leading light spot. The corner window is extended with desk and joined with wash basin. At the turning is fixed a drop-shaped lighting, boasting the cohesion of the corner.

Outside the master bedroom is a greenery world. Inside it, comes the view of 101 Building, a landmark of Taibei. The bedhead is decorated with leather and Swarovski Rhinestone. The black and the white set off the fashion, luxurious but not very.

Newly Born Churn to Collet Home Warmth

教堂新生，收藏家的温馨

设计公司：家镁室内设计
设计师：家镁
摄影师：约亚斯·苏扎

Design Company: Gianna Camilotti Interiors
Designer: Gianna Camilotti
Photographer: Joas Souza

本案室内设计于2014年完成。除了楼上新添白色木质地板来代替旧时地毯，设计未对原建筑特色、饰面作任何改动。些许家具铺陈由本案设计师自行设计，如休息室里的大沙发、靠垫、贴画、餐桌等等，大部分现代化的家具、灯具都出自意大利名师之手。设计紧随主题，竭力创造出一个对朋友开放，既可以享受美食又可以K歌的欢乐空间。该空间同时具有伦敦钢琴吧的气氛。

Gianna Camilotti Interiors has done its Interior design in 2014. All architectural features and finishing have been kept untouched, apart from the carpet that has been replaced for white wooden floor (upstairs). Most furniture and lighting are contemporary pieces by Italian designers, but some have been designed by myself (as the large sofa in the lounge, cushions, posters and dining table). From the start designer had in mind a house open to friends, to enjoy cosy dinners and karaoke parties that had an atmosphere of London piano-bars.

一层平面图 / First Floor Plan

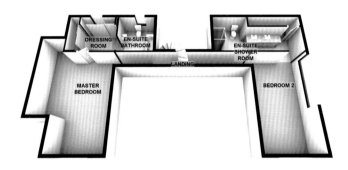

二层平面图 / Second Floor Plan

Atrium Life: Half-Recreational, Half-Scenery

一半娱乐一半风景，轻松生活向中庭敞开

项目名称：高原别墅
设计师：克劳迪娅·赖斯
摄影师：尼尔森·科恩
面积：600 ㎡

Project Name: Planalto House
Designer: Claudia Reis
Photographer: Nelson Kon
Area: 600 m²

"高原别墅"量体呈直角棱柱式，与所在街面垂直。上面空间隐密性特好，整个空间只占所在地基面积的一半，另外一半完全让位于娱乐、风景。

平行于街面的金属立柱依边缘排列，强化着各独立量体的地位，同时揭示着整个别墅的结构功能。挑檐之下是主入口平台。1.8米宽度的廊道游弋于整个空间，连接着各功能区域。室内轩亭、烧烤区域是其中的两个关键分区，可以用作社交等娱乐场合。

车库与娱乐区的上面，设有一个花园式阳台。此处是个多功能空间，借助于娱乐区的楼梯便可轻松到达。

不同区域因其使命不同，用材而各有特色。有的使用透明材质，有的则完全起到遮挡视线的作用。空间道道混凝土墙起着分割、界定的作用，而大大的滑动门则引入风景。

城市居家生活，子女绕膝，其乐融融，尽在本案"高原别墅"之中，好一个现代的巴西建筑空间。

This is a rectangular prism, perpendicular to the street and contains the intimate features of the house on the upper floor, occupying only half of the land and releasing the other half for recreation and landscaping.

The metal beams on the edge of the volume parallel to the street, reinforce the idea of independence between the volumes and reveal the structural functioning of the house. The main access platform, located under the front overhang on the main floor, provides access to the corridor 1.80 meters wide running through the house, connecting various environments. After passing through the service area, we come to the point of access to two key areas: the social rooms (like an indoor pavilion) and the barbecue area (recreation).

A garden-terrace covers the main floor block of the garage and recreation area. It can be accessed by the stairs at the recreation area. It is a space of multiple functions. The characteristics of the materials used in this residence as chromaticism, texture and transparency were carefully chosen because of the intentions pursued in each space. While the transparence integrates, the concrete do the opposite. The concrete walls divide the space, while the large sliding glass doors bring the landscape into the house.

The Planalto house was conceived as an urban house for a couple with children and could be considered as exemplary of the current Brazilian contemporary architecture.

现代酷宅 ● 109 ● Modern Cool Mansion

美式阔宅
American-style Mansion

Time Brews the Flavor of Life

用时间酝酿生活的醇香

项目名称：海点别苑
设计公司：罗伯特建筑师事务所

Project Name: SeaPoint
Design Company: Robert Hidey Architects

"海点别苑"共有571平方米,"托斯卡纳"风格,内设5个卧室、4个车库。室内外空间尽享艺术生活。除了中央庭院,空间设有凉廊、室外日光浴场所、遮挡阳台、泳池、SPA、室外厨房。家居生活有太平洋的浩瀚风景相伴。

"法式门"、"C"型平面图无缝衔接庭院与家庭娱乐空间。沿着"C"字的起点走到底点,有一餐厅,开向院落,正对着院子中光亮的壁炉。大房间、客厅分列两边。厨房里除了Sub Zero品牌家用电器,还有包裹着石材的橱柜,尽显空间特色。紧邻着正式餐厅的是一个管家"餐具"室,以备不时之需。气温得以控制的酒窖,同样石材饰面,传统的意大利设计,水平排列着的酒架,摆放着东家最为看重的家藏。米黄色的正餐厅与酒窖实现着无缝对接。身处其中的客人,感受的是一种亲密的气氛。主卧附设起居室、男女主人步入式衣橱。静卧于隐密之中,更可尽享太平洋的波澜壮阔。楼下的次卧还有一个后撤式室外休息室。

太平洋的大美壮观,令人惊叹的美学设计,为客人提供了一个经典的意大利世界。

SeaPoint, a 5-bedroom, 4-car-garage, 571-square-meters formal Tuscan-style home reflects a relaxed approach to artful, outdoor living with inviting exterior areas, such as a central courtyard, sheltered loggias, and an outdoor sundeck. A covered patio, pool, spa, and an outdoor kitchen capture the home's spectacular Pacific Ocean backdrop.

Flowing naturally to the exteriors through French doors, SeaPoint's C-shaped design allows seamless access to the courtyard from the home's entertaining spaces. At the foot of a C-shaped pavilion, the dining room looks out onto the courtyard's illuminated fireplace, while the great room and living room flank the space. The kitchen, replete with Sub Zero appliances including warming drawers, also features counters clad in stone. Off the formal dining room is a butler's pantry for additional preparation space. The stone-clad wine cellar, traditional in Italian design, features horizontally spread wine racks to display the owner's prized bottles in this climate-controlled cavern. A soothing beige-toned formal dining space effortlessly connects with the cellar offering an intimate space for guests to retreat. The master suite with a sitting room and his-and-hers walk-in closets showcases spectacular views of the Pacific Ocean. Downstairs, the junior-master suite offers a private outdoor retreat.

Seapoint's romantic hillside location with Pacific Ocean views and stunning, authentic design transports visitors to a classic Italian world.

浓墨重彩调制美式优雅

American Grace by colors

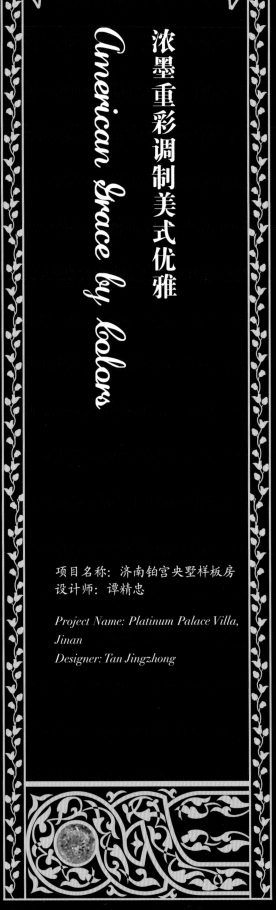

项目名称：济南铂宫央墅样板房
设计师：谭精忠

Project Name: Platinum Palace Villa,
Jinan
Designer: Tan Jingzhong

本案定位为美式风格，大量橡木护墙板的运用，不仅让空间更为连贯，更营造出沉稳、典雅与怀旧的空间气氛。几乎每个区域，设计师都选择了色调浓重、华丽的装饰，行走其间，如同置身于文艺复兴时期的艺术气息所包围的小宇宙当中，接受艺术的洗礼与熏陶。

客厅大部分墙面被棕褐色的橡木护墙板包裹着，色泽典雅、纹理美观，给人以高雅、雍容的感觉。帘幔、地毯和家具、陈设饰品也均以华丽、高贵为主，充分体现了美式新古典的特征。站在楼上俯看小客厅，古典缎面的家具、图案丰富的地毯与镜面元素交织，在水晶吊灯的照耀下愈发熠熠生辉。餐厅黄色花纹的布面餐椅和优雅华丽的窗帘延续了客厅的精致感，具有一种戏剧化的华美和绚烂。厨房橱柜虽仍以棕褐色橡木打造而成，然而地面却运用米白色大理石进行铺贴，大大降低了深色橱柜带来的厚重感。

走进书房，天花板设计独特，给人眼前一亮的感觉。从墙面到书柜、书桌，设计皆以橡木打造而成，四处弥散着木材的高贵、复古的气息，不但给人雍容之感，更彰显出主人的气质和修养。此外，这里还特别设置了会客区。书香萦绕的小空间里，小巧的古典沙发与茶几静静地躺着，在这里跟亲朋交谈，别有一种情趣。

A project this space is positioned as American style, where amounts of oak chair rail not only make the space more continuous but bring forward a nostalgic ambience of staidness and grace. Almost all sections are fixed with luxurious decorations of heavy hues, allowing for feelings as if you were surrounded with an artistic sense of Renaissance to receive baptism and edification of art.

Elegant in color, beautiful in aesthetics, the chair rail in the living room offers grace and elegance. Window curtain, carpet, furnishings and accessories focus on magnificence and dignity. All fully embody characteristics of neo-classical American style. Furniture coated in classical brocade, patterned and mirror are interspersed and glister with light cast on by chandelier. Fabric dining chairs and curtains keep the delicate sense. That's a dramatic brilliance. In the kitchen, the off-white marble flooring successfully grounds off the stiffness by cabinet of brown oak.

In the study, the ceiling looks refresh. Surface blocks from walls, to bookshelf and to desk are all made of oak, oozing a noble and retro sense, not only providing grace and elegance, but embodying personal taste and culture. Additionally, there is an area especially for receiving. In such a setting with book air, chatting on classical sofa accompanied with sofa makes fun and interest of another kind.

Breathe Simultaneously with Nature

海天一色，与自然同呼吸

项目名称：长岛别墅
设计公司：1100 建工
摄影师：尼古拉斯·柯尼希
地点：纽约长岛

Project Name: Long Island House
Design Company: 1100 Architect
Photographer: Nikolas Koenig
Location: Long Island, New York

美式阔宅 ● 127 ● American-style Mansion

剖面图 / Sectional Drawing

本案位于长岛东岸，其设计如同自然山水
天生一般。空间量体分为主体别墅、亭台两个
分。沿着北部车道前行，别墅气势宏伟，石材
面，长方形大窗方正规整。南立面用玻璃全副
装，伴着阳光，静享大海风情。现代化的盒体
无意之中成了一个开放的雕塑造型。

沿着坡形的绿色屋顶，绵延起伏的沙丘与
地的植被，强化了别墅从自然中而生的意象。
墅四周的绿植，都是当地"土著"，进一步强
了建筑与自然山水的合二为一。同时，石材色
的立面以其色彩呼应着沙滩。海洋色调的涂装
利于建筑抵御暴风雨的侵袭。

建筑北面看湖，南面观海。湖是淡水湖，
是蓝色的海。进入主体空间，首先是一处玻璃
护的区域，采光的同时，更让视线越过湖水，
过沙丘，直入大海。生活区、客厅、厨房簇拥
倾斜的屋顶下，使上、下空间连接的同时，更
下层空间向双高的空间洞开。

Situated on the eastern shore of Long Island, the de
of this beach house is conceived as emerging from
landscape. The house is composed of two elements –
main house and a pavilion – that are situated on a
that extends into the landscape. From the drive
entrance on the northern side, the house appear
a largely solid, stone-clad volume with rectangu
perforations for the windows. In contrast, the n
private southern façade is composed almost entirel
glass to maximize ocean views and sun exposure.
initially presents itself as a classic modern box unfole
become an unexpectedly open sculptural figure.

The dune landscape and native plantings along
ramped portion of the green roof emphasize the vi
metaphor of the house springing forth from the terr
Native plants surrounding the house recreate the s
natural conditions so that the building is truly integr
with the landscape. Reinforcing this concept, the ston
the façade matches the color of the beach sand. In add
to the robust stone cladding, marine grade finishes
used to make the exterior of the house resistant to storr
The house is in a unique position to capture views of
ocean and a freshwater pond. The design optimizes t
views by angling the north façade toward the lake
aligning the south façade with the ocean. Upon ente
the main house, visitors encounter an open common
enclosed by glass walls that let in an abundance of nat
light and provide unobstructed views out to the inf
pool, across the dunes, and to the ocean beyond.
living, dining, and kitchen spaces are drawn togethe
the sloped roof, which creates continuity by connecting
first and second floors, and opening up the first floor
double height space.

Facing the Sea in Spring

面朝大海，春暖花开

项目名称：长岛别墅
设计公司：BM 建工
摄影师：马克·布莱恩·布朗

Project Name: Long Island Villa
Design Company: Blaze Makoid Architecture
Photographer: Marc Bryan-Brown

长岛别墅位于长岛东岸。滨海前沿，占地4 046平方米，地基狭长。2010年联邦新修订"建筑水位"法规要求一楼空间高度要超出海平面5.18米以上，最大空间高度为12米。另外，空间位于高速风带之内。一切所有，加大了建筑平面与结构的难度。

建筑量体给人一种清爽、直白的感觉。两层的入口玄关只设有一个开口，另有一悬臂式楼梯势如拔地而起。沿墙是各服务功能区域。向里走，才是滨海前沿。那里设有开放式的生活区、餐饮区、厨房。4.57米的落地推拉门最大化地引入海天美景。穿过开阔的门，便可到达外面的庭院与泳池。

一楼的设计，如同"玻璃层"。二层的设计却如同用"洞石"、"玻璃"打造的抽屉，悬于"玻璃层"的上空。自西向东三个大小相同的儿童卧室，中间夹以主卧。主卧的阳台，如同墙体的凸出物。现场浇筑的大理石板，加尔各答的大理石面板，非洲红豆木的预制木质材料强化了空间的优雅与线条的美。

Sited on a narrow, 4,046 square meters, oceanfront lot on the Eastern Shore of Long Island, the design of this house was one of the first projects in the Village of Sagaponack to be affected by the 2010 revision to FEMA flood elevations, requiring a first floor elevation of approximately 5.18 meters above sea level with a maximum height allowance of 12 meters. The location within a high velocity (VE) wind zone added to the planning and structural challenges.

Makoid wanted the structure to appear simple and clean upon arrival. The two storeys travertine entry facade is highlighted with a single opening accentuated by a cantilevered afromosia stair landing that hovers off the ground. A layer of service spaces run parallel to the wall plane creating a threshold prior to reaching the horizontal expanse of the open plan living room, dining area and kitchen that stretches along the ocean side of the house. 4.57 meters wide floor to ceiling glass sliding panels maximize the ocean view and create easy access to the patio and pool beyond.

The second floor is imagined as a travertine and glass "drawer" floating above the glass floor below. Three identical children's bedrooms run from west to east, setting a rhythm that is punctuated by a master bedroom with balcony that projects from the wall plane. It is clad in the same afromosia wood as the stair landing. The quiet elegance and clean lines of the house are accentuated by the materials that also include poured-in-place concrete floors, Calcutta marble cladding and afromosia millwork.

Natural Roughness and Life Delicacy Fused into Idyllic Pastoral

自然的粗犷与生活的细腻融合成田园牧歌

设计师：Ozhan Hazirlar
网址：www.ozhanhazirlar.com
邮箱：ozhan@ozhanhazirlar.com

Designer: Ozhan Hazirlar
Website: www.ozhanhazirlar.com
E-mail: ozhan@ozhanhazirlar.com

別墅起居室由 Ozb
Hazirlar 担纲设计。
是一个概念性项目，
土耳其比尔肯特的一
别墅的一部分。设计
合了古典和现代两种
格。

室内采用了暖色
装饰，并且采用舒适
人力工学家具。这座
墅总占地面积为 320
方米，而这一舒适的
住空间面积为 60 平
米。这个宽大的居住
间与花园相连，彰显
设计的优雅与品质。

Villa Living Room Pro
has designed by Ozl
Hazirlar. It is a conc
project of a part of vill
Bilkent neighborhood
Ankara where capito
Turkey. Classic and mod
style are integrated in
design.

Warm style of colors
used. Aim to use ergono
and comfortable furnit
This wonderful living p
have 60 square meters are
total 320 square meters w
house. This huge living a
combined with garden h
great design reflects elega
and quality.

Return to the Innocent of American Manor Life

美式庄园生活的返璞归真

项目名称: 苏州九龙仓国宾一号别墅样板房室内设计
设计公司: 上海大研室内设计工程有限公司
设计师: 欧阳辉
面积: 375 m²
用材: 天然石材、仿古砖

Project Name: Show Villa, Number One State Guest, Wharf, Suzhou
Design Company: Shanghai Dayan Interior Design Institute
and Engineering Co., Ltd.
Designer: Ouyang Hui
Area: 375 m²
Materials: Natural Marble, Antique Brick

本案位于苏州工业园区金鸡湖大道旁，位置十分优越。此套样板房定位为美式古典风格。在室内环境中力求表现悠闲、舒适的生活状态。运用了天然石材、仿古砖等材料来体现业主对优雅的生活追求。

一层设有客厅、餐厅和厨房。中、西式厨房的巧妙结合，满足了主人对美食的热爱以及对生活品质的追求。客厅与餐厅中，纯色铁艺的材质体现出美式风格的粗犷与自然感。二层是主人的私密空间。在美国人的价值观念中，卧室应该是最豪华的地方。本案中，进门处是独立的书房，西边是主人的卧室、更衣间和卫生间。整个空间的线条流畅且功能统一。三层是儿童房。男女小朋友都有自己的主题生活空间和独立的活动区域。地下室部分主要设置了视听室、书房等多功能空间，体现出主人对生活品质的追求。

该案给人以原始而简洁、粗犷而随性的感觉，满足了人们返璞归真的心理需求。

This apartment is located in and ideal place, beside Jinji Lake of Suzhou Industrial Park Avenue. This setting of model room is positioned American classical style. The indoor environment is designed to be with relaxing and comfortable living conditions. The use of natural stone, antique tiles and other stones reflects the owners' pursuit of the elegant life.

There are living room, dining room and kitchen on the 1st floor of the villa. The unique combination of Western-style and Chinese-style kitchen meets the owner's love for food and the pursuit of quality of life. The solid iron material in the living room and the dining room reflects the rough and natural quality of American style. The 2nd floor is the master's private space. In American values, the bedroom should be the most luxurious place. In this apartment, an independent study is set right facing the door, and the master bedroom, the dressing room and the bathroom are to the west. The entire apartment is endowed with smooth lines and unified features. The kid's room is on the 3rd floor. The boy and the girl have their own living spaces and activity areas. The audiovisual room, the study and other multi-functional spaces are housed in the basement, which reveals the owner's pursuit of quality life.

The design presents a style primitive and simple, rugged and casual, which satisfies people's psychological needs to return to nature.

Spatial Broadness and Openness Scented with Wood

开阔空间，处处盈溢木质清香

项目名称：大美别苑
设计公司：大卫建筑室内设计
设计师：大卫·格拉
地点：巴西贝洛哈里桑塔
面积：350 m²

Project Name: Apartment Lal
Design Company: David Guerra Architecture and Interior Design
Designer: David Guerra
Location: Belo Horizonte, Brazil
Area: 350 m²

该空间集城市、乡村别墅风格为一体，美观性与实用性兼具。为了满足业主夫妇及两个孩子的生活需要，拆除了客厅及阳台的隔离墙，整个空间因此显得更为开阔、流畅、舒适。阳台成了可以兼用餐、厨房的另一空间，适宜朋友小聚、家庭早餐，同时可远观山景。阳台配有亚麻沙发、椅子、古式扶手椅，因此阳台俨然又成一个适宜放松的生活区。客厅的壁炉虽小，但却成为客厅的主要特色。温暖的气氛里散发质朴、自然的气息，现代的材质融汇着工艺感十足和现代气息。不同色彩的羊毛、象牙、皮材、旧木、石头、不锈钢、黄色的金属、钢、镜材、玻璃、亚克力等，共同打造出一个极好的环境。

家具铺陈无不是为了空间的舒适、温暖、优雅及放松。

美式橡木餐桌彰显着皮质椅、吊灯的美丽、轻盈与舒适。

整个空间的地板，除了洗浴区，全部以红木铺就。宽广的红木板采购于农场。经过漂白，木板极好地保持了原木的质朴，整个空间也因此显得更为明亮、现代。

厨房使用多种用材，石墨色、银白色的赛丽石分别铺于地板、台面，墙体施以黑色、灰色的液压砖，木门、红木桌由建筑师亲自设计。整个厨房因此显得舒适而现代。卫生间延续着厨房用材的混搭风格。地板、台面同样采用透明的赛丽石。深柴色的德国汉斯格雅卫浴与墙砖、镜子形成鲜明的对比。

主浴空间运用了品牌石灰岩，主打优雅。浴柜镜面与柜门为意大利品牌，不失豪华手笔。

主卧室里的床头板由红木、不锈钢制作，梳妆台是法式的。各种用材保持着固有的天然色调，加上亚麻、皮革、实木，为主人提供了一个放松的场所，恭候着主人的归来。

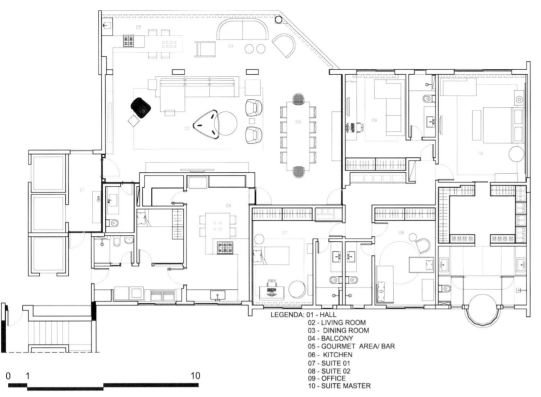

LEGENDA: 01 - HALL
02 - LIVING ROOM
03 - DINING ROOM
04 - BALCONY
05 - GOURMET AREA/ BAR
06 - KITCHEN
07 - SUITE 01
08 - SUITE 02
09 - OFFICE
10 - SUITE MASTER

0   1   10

平面图 / Site Plan

The new home combines the coziness aspect of a country house and the urban and practical style of the big city. To attend the needs of the couple with two children, a renovation was needed. The walls that divide the living room from the balcony were demolished to combine the ambient with larger, fluid and comfortable space. The balcony became a gourmet bar/kitchen that can be used for the wine with friends and breakfast in family with a view of the mountains. Linen sofa and chairs and a vintage armchair appear as a relaxing living area also in the balcony. A small fireplace has become a major element of the living room wall. The new warming ambience mix colors, rustic and natural materials with modern and technological ones. They are wool, natural linen, nude tones, leather in different colors – honey, whisky and chocolate, woodland demolition wood, grey Mister Cryl, Silestone rock, stainless steel, yellow metal, bronze, mirror, glass and acrylic, all materials that combined, gives a great ambience.

The choices of the furniture, noted the concern of creating a place that prioritizes comfort, warmth, elegance and relaxation. The dining table with an American toned oak highlights the

beauty, lightness and comfort of the Tombly leather chairs from Minotti and also the chandelier by Mooi.

The entire floor of the apartment, except the wet areas, had been replaced by wide planks of mahogany field bought from a farm. The floor has gone through a bleaching process, maintaining the identity and rusticity from the wood and giving a more light and modern touch to the place.

The kitchen also provides a mix of materials, the technology of Italian glass Panna and reflective glass, Italian chairs Papiro by B&Bitalia, graphite Silestone on the floor and silver one on the countertops, walls with black and grey hydraulic tiles, wood doors and mahogany table –design by the architect. The kitchen becomes a mix of cozy and contemporary at the same time. That mix can also be seen in the toilet with gauzy Silestone floor and countertops, burgundy Mister Cryl, Hansgrohe mixers that contrast with the tile of the wall and the Indian mirror.

In the master bathroom, the priority was the elegance, which was achieved by the Limestone Persiano, cabinet with Italian glass and Rimadesio door.

In the master bedroom, the highlights are for the headboard with mahogany with stainless steel profile, French dresser, linen Selene bed by Maxalto and Pantosh wooden chair. Nude and caramel tones and natural materials, linens, leathers and woods, provide a welcoming place that facilitates relaxation.

美式阔宅 ● 149 ●

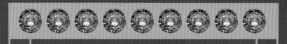

亲近自然，阅读四季

Be Close to Nature to Read Seasons

项目名称：枫桦别墅
设计公司：理查德·科尔建筑师
事务所
设计师：理查德·科尔、卡拉·威
尔福德
摄影：西蒙·伍德
面积：591 m²

Project Name: Angophora House
Design Company: Richard Cole
Architecture
Designer: Richard Cole, Karla Wilford
Photographer: Simon Wood
Area: 591 m²

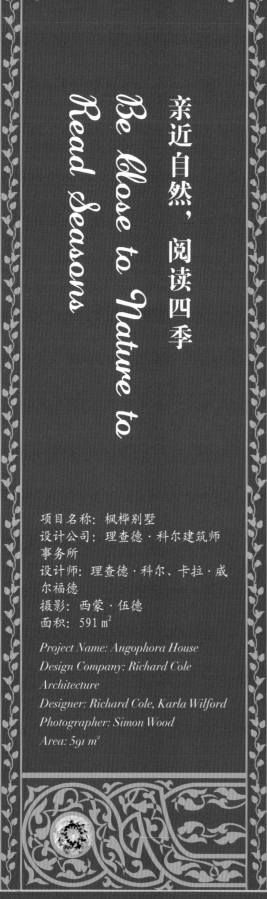

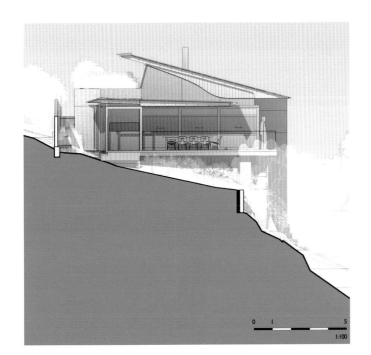

剖面图 / Sectional Drawing

剖面图 / Sectional Drawing

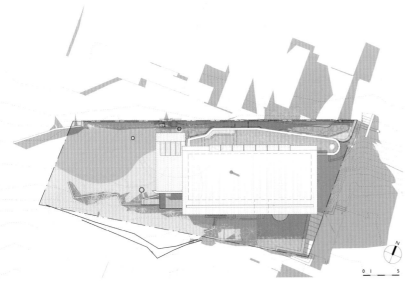

平面图 / Site Plan

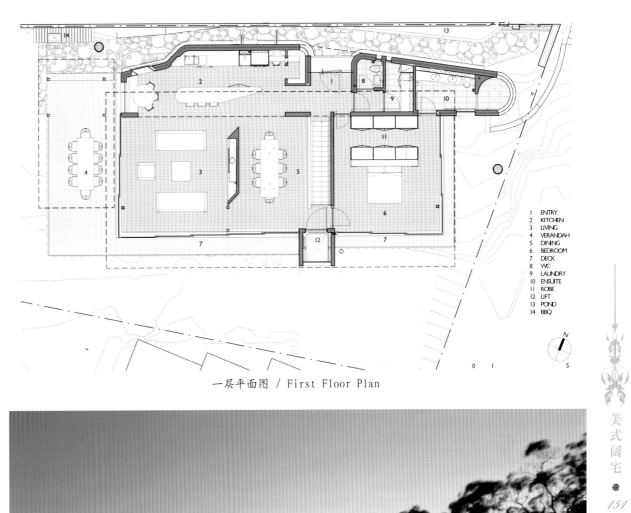

1 ENTRY
2 KITCHEN
3 LIVING
4 VERANDAH
5 DINING
6 BEDROOM
7 DECK
8 WC
9 LAUNDRY
10 ENSUITE
11 ROBE
12 LIFT
13 POND
14 BBQ

一层平面图 / First Floor Plan

　　枫桦别墅的业主为一对老年夫妻。其儿子即是本案建筑师。本案旨在于内敛、不事张扬的背景下，通过自然的用材与质感，为业主打造一个出入自由、温馨的空间。颐养天年的居所，极尽美丽。虽然工艺考究，却对其制作痕迹毫不掩饰。

　　人老了，终归有所留恋。于是设计妥善的安排了业主一生的收藏。无论是威士忌酒瓶，还是厨房的旮旯角落，抑或是阳台，一切都是那么井井有条。既是可以避难的圣殿，也是可以享受的画廊。所有一切皆归功于建筑师对其父母的理解与了解。Artek品牌的桌椅，是20世纪60年代置办的。空间内当然不能少了建筑师本人为其父母设计的黑木桌，同时理查德·沃恩设计的餐柜也悄悄地立于一旁，相互呼应。

　　本案位于悬崖之边，所处地区人口稠密，为城市化文物保护地区。自然空间内便有了洞穴、平台、树冠等可以彰显地域色彩的元素。进入空间，如同开启一段旅程。宁静的庭院内，开挖着一个线性池塘，立着一堵石墙。北面是混凝土、砖石结构窑洞般的房间。雕塑质感华盖般的屋顶下是可以欣赏到山谷景色的房间。房间有着坚实的墙。两个悬空的平台凌空紧靠，直面着陡峭的岩壁。

　　空间设计紧凑而不显空阔，极好地满足了老年人的心理。所涉用材，包括混凝土、钢结构、滑动木窗、屏风、旋转门、木质天花板、木工、包锌、钢窗、无框曲面玻璃、隐蔽照明、石雕等等全部手工定做，尽显工艺之精湛。作业时，予经济元素及质量元素同样之考量。相比众多现代的房舍，本案写不尽的内敛、低调，始终是首要考虑因素。建筑设计时，本案进行了大量的岩土工作，特别是车库的挖掘，代价极大。选材主要考虑质量、耐久性和抗风化。外部木材使用极加限制，以降低维护成本。建筑简单、自然，但却细致、美丽、丰富并且耐用。

Built for the architect's parents, the house is fully accessible and designed for "ageing in place". Of conservative background, the design of their final house is the embodiment of a leap of faith. Natural materials and textures achieve a raw warmth. The form is carefully composed and articulated with beauty the objective. Meticulously crafted, it does not conceal the marks of its making.

The house acts as a container for the possessions accumulated over a lifetime. The architect knew intimately how his parents lived, their habits and routines. There is a place for a whisky bottle, a kitchen nook where busy schedules can be organized, a deck where dusk can be savored. The house is both sanctuary and gallery. The Artek breakfast table and chairs are the same treasured design as purchased in the 1960's. The Blackwood table designed by the architect complements an existing Richard Vaughan sideboard and responds specifically to the space it occupies.

Built over an escarpment in a densely urbanized heritage conservation area in Waverton, the form responds to the difficult site using elements of cave, platform and canopy. The design anticipates a sequential experience: entering from the street to a tranquil courtyard with a linear pond and stone wall, though the concrete and masonry cave-like northern rooms, to the light filled open space beneath the sculptural canopy roof, a room with the refuge of encompassing solid walls and the prospect of views over the valley. Two platforms launch from the anchoring escarpment. The unexpected raw rock face is revealed below.

The brief called for a compact but spacious feeling light filled house to accommodate the needs of a couple as they aged. The house is a testament to the dedicated craftsmen who built it. The concrete, steelwork, sliding timber windows and screens, pivot doors, timber ceiling and cladding, joinery, zinc cladding, steel plate windows, frameless and curved glazing, concealed lighting and stonework are all custom made hand crafted elements. While overall economy was always an important design parameter, quality was always the primary consideration. Economy of scale was an objective, and the scale of the project restrained when compared to many contemporary houses. The difficulty of the site and restricted access *options necessitated significant and expensive geotechnical work, particularly to excavate the garage. Material selection focused on quality and durability. Weathering is considered. Timber is limited externally to reduce maintenance. Richness is achieved through the simple detailing and beauty of natural materials. Inherent value is achieved by longevity.*

南非华宅

South African-style Mansion

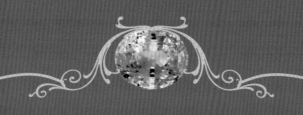

Free or Recessive to Create Life Possibility

收放自如，创造生活无限可能

项目名称：DW34
建筑公司：SAOTA 建筑师事务所
建筑师：斯蒂芬·安东尼、博比·赫吉尔、
杜克·威廉姆斯
设计公司：OKHA 室内设计
设计师：亚当·考特
摄影师：亚当·莱奇
地点：南非开普敦

Project Name: De Wet 34
Construction Company: SAOTA - Stefan Antoni Olmesdahl Truen Architects (www.saota.com)
Architect: Stefan Antoni, Bobby Hugill, Duke Williams
Design Company: OKHA Interiors (www.okha.com)
Designer: Adam Court
Photographer: Adam Letch
Location: Cape Town, South Africa

本案整个临街的立面只用了两种材质进行装饰。一种是周围空间常见的泥板岩，另外一种为美国红杉木。美国红杉木是一种可持续利用的材质，不易被自然风化，也无需任何表面装饰。客厅地板所用的花岗岩板材，边角手工打磨，给人一种古老的质感。

客厅也因此与整个空间有一种细微的差别，更显精致。水景处，排列着来自中国福建东北部的花岗岩纹理石砖，如同一个脊柱把房间两翼紧密地连接在一起。客厅西部，玻璃釉面，便于观景。东部呈"L"形，与轻盈的西部相比，显得更为厚实。客厅区域还设有鸡尾酒吧，酒吧由两块巨大的鹅卵石构成。两块石头重达5 000千克，几经切割、打磨，最终成形，并以吊车运至此处。

厨房与家庭室位于庭院花园。花园一角有一篝火坑。夏季的室外，无疑是吸引目光的消夜之所。

The entire street facade is made up of only 2 materials, rough dark grey shale stone found in the immediate vicinity and American redwood, a sustainable wood, known for its ability to weather naturally to a light greyish color without any surface applications.

The arrival in the Living Room is marked by the subtle change in floor finish from the overall floors which are Neo Sardo granite with a hand hewn soft edge giving it an ancient quality. A Silver Grey veined granite tile from Fuzhou in the North-east part of Fujian Province, China, lines up with the water feature creating a spine running through the house connecting the two wings, the double volume glazed Living sea facing wing on the left west side and the more solid and sculpted L-shaped wing on the east overlooking the courtyard garden. A massive rough-hewn granite boulder (made up of two huge natural boulders and painstakingly torched and chipped to get to the final shape. It weighs 5,000 kilograms and needed to be brought into position by crane) forms the cocktail Bar at the arrival to the double volume Living area.

Situated around the Courtyard garden are the Kitchen and Family Rooms. An outdoor Boma (African campfire enclosure) is situated in the corner of the Courtyard garden. This is the focal point of summer outdoor living.

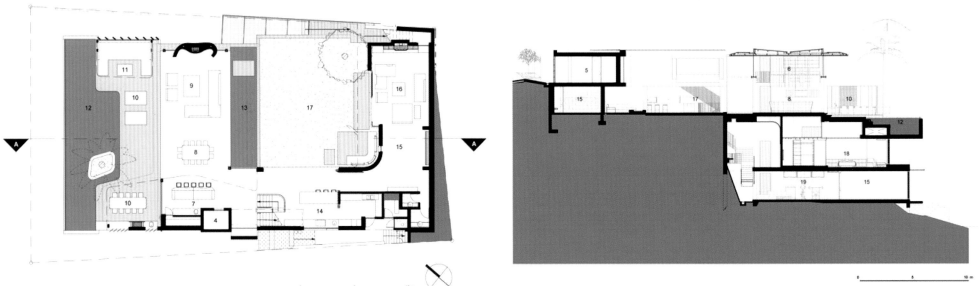

一层平面图 / First Floor Plan

剖面图 / Sectional Drawing

To House Hills and Waters, to Enjoy Life Leisure

藏山拥海，阔逸人生

项目名称：山海花苑
建筑公司：SAOTA 建筑师事务所
建筑师：格雷格·图恩、斯蒂芬·安东尼、特斯维尔·萨斯
室内设计：奥卡室内设计
摄影师：亚当·莱奇

Project Name: La Lucia
Construction Company: SAOTA – Stefan Antoni Olmesdahl Truen Architects
Architect: Greg Truen, Stefan Antoni, Teswill Sars
Interior Design: OKHA Interiors
Photographer: Adam Letch

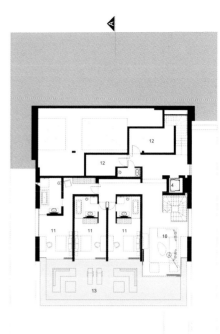

一层平面图 / First Floor Plan

二层平面图 / Second Floor Plan

三层平面图 / Third Floor Plan

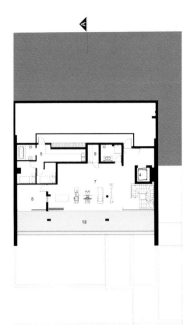

地下层平面图 / Basement Plan

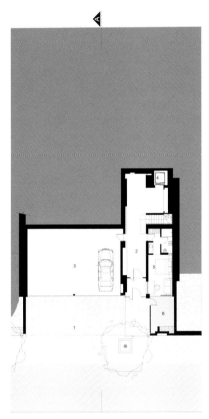

入口平面图 / Entrance Plan

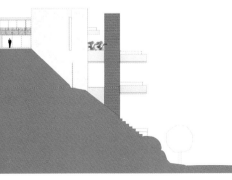

立面图 / Elevation Drawing

正如其名所示，山海花苑极尽山海之景，但同时最大化减少周围建筑对其的影响。基地位于"狮子头"山巅，西面波澜壮阔的大西洋。

双量体空间连接着东部花园，同时充分利用着西部的海景。玄关动感十足，以垂直态势，呼应着生活起居空间。卧室层垂挂于屋檐，如悬似漂。除却卧室层，另有两层空间。

柔和的S曲线，囊括着主卧空间。附属的空间，以系列核桃百叶窗独立于生活空间。宏伟的内部空间，有戏剧性张力与动感的旋律。室内装饰，柔和亲切中有着对立，但却不影响纯净、几何性质的线条架构。

不同的质感与饰面，是自然、有机、舒适的体现。共存之间，书写着的依旧是宁静与祥和。

"The brief was to create a building that maximized the connection with the view and the sea, create a garden on the mountain side of the site and to minimize views over neighbouring buildings."says Greg Truen, Project Partner. The site is positioned on the side of Lion's Head and has fantastic views over the Atlantic Ocean to the west.

The decision was to create a double volume space at the top of the site that could connect the garden court to the east, while also taking advantage of the views over the ocean to the west. The entrance hall is a dramatic, vertical space and provides a counterpoint to the living spaces. The volume of this large area is modulated by the floor which has two levels, and by the bedroom floor which floats into the space and is hung from the roof.

This element, which has a soft s-curve, contains the master bedroom and its ancillary spaces are separated from the living room double volume by a set of walnut shutters. "The building provides a volumetrically dramatic and dynamic interior space on a grand scale. The objective of the interior décor was to create a softer and intimate counterpoint whilst not jeopardising the clean, geometric lines of the architecture. These elements can harmoniously co-exist and work off each other." says Adam Court of OKHA Interiors.

By utilising a broad base of textures and finishes, the décor feels natural and subtly organic, comfort being of paramount importance at all times; the overall ambiance is one of calm and serenity.

Dwelling with Water Around Makes Life a Big Party

绕水而居，生活就像大 Party

项目名称：霍顿 1448
建筑公司：SAOTA 建筑事务所
建筑师：格雷格·图恩、艾娜·傅立叶
设计公司：安东尼联营公司
设计师：马克·里利、萨丽卡·雅各布斯
摄影师：亚克·莱奇、艾尔莎·杨
地点：南非约翰内斯堡

Project Name: 6th 1448 Houghton ZM
Construction Company: SAOTA - Stefan Antoni Olmesdahl Truen
Architect: Greg Truen, Ina Fourie
Design Company: Antoni Associates
Designer: Mark Rielly, Sarika Jacobs
Photographer: Adam Letch, Elsa Young
Location: Johannesburg, South Africa

整个建筑空间由两个量体构成。两个量体共有一个入口，共用一条车道。车道位于南部。两座房子皆以"U"形配置组织在一个内置庭院的周围。

双高的隔离墙介于公共的前院与后面隐私区域之间，起到很好的界定作用。水平的沟槽，便于白天日光的渗透及夜晚室内灯光的流出。东面的建筑地势高于西面，一主一次的印象油然而生。

室内空间的设置围绕正中央的造型进行。造型上披有一个绸带，绸带有点螺旋楼梯的模样。向前走，是客厅区、厨房及北部的私人花园。西面是娱乐室，娱乐室附近设有泳池及健身房。所有这些空间都可通向庭院，庭院里绿树掩映，别有景致。

因为地基的布局，空间建筑量体都有一个朝西的立面。立面装有大大的百叶窗，百叶窗垂至一楼水泥板，遮挡着外面的太阳。

卧室位于上层空间。子女房、书房面朝北部。父母房、客厅位于双高空间的另一端，面朝西、北两个方向。沿着开口的墙有一条走道正好通往子女房。透过墙上的开口，倾泄进道道光束，原本平常的空间顿时有了一种魔幻般的感觉。

The site was split into two separate sites organized around a common entrance and driveway to the south which provides access to the two houses. Both houses have a U-shaped configuration allowing them to be organized around an internal courtyard.

A double height wall separates the public forecourt from the private domestic spaces. It is perforated with a series of horizontal slots that allow daylight in during the day and artificial light out at night providing glimpses of activity in the house. The house to the east is set higher than the house to the west affording it views over its partner.

Internally the spaces pivot around a central volume with a ribbon like spiral stair. Beyond this element are the living spaces and the kitchen with a private garden to the north. To the west is a games space with a pool and a gym. All of these spaces connect to the courtyard which in turn connects back to the main house and its living spaces. Trees have been positioned in the courtyard to create a green centre to the composition.

Because of the configuration of the site, each house has a substantial west facing façade. A set of large shutters, which drop below the level of the first floor slab, provide shade and protection from the setting sun. Care was also taken in selecting performance-glass that would minimise the impact of direct sun.

The bedrooms are on the upper level; the children's bedrooms and a study space facing north and the parent's bedrooms and a living room on the other side of the double volume facing west and north. The passage to the children's bedroom runs along the perforated wall that separates the building from the driveway. The perforations allow light in creating a magical space from what would normally be quite mundane.

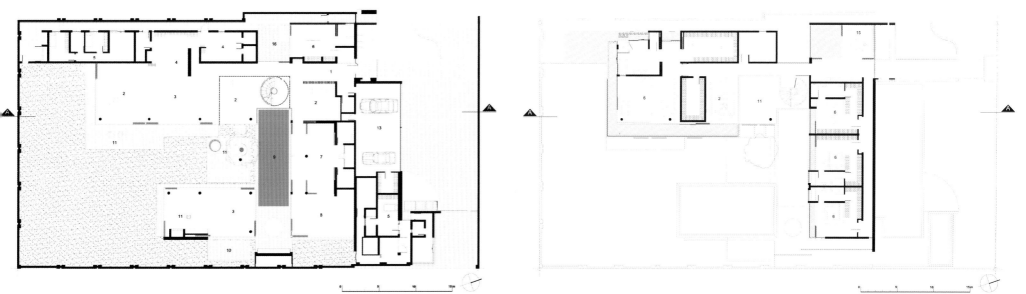

一层平面图 / First Floor Plan

二层平面图 / Second Floor Plan

Water Surrounding, Transparency Natural

环翠拥水，通透自然

项目名称：珍珠谷之"环翠拥水"
设计公司：安东尼联营公司
设计师：马克·里利、乔恩·凯斯、萨丽卡·雅各布斯
摄影师：亚当·莱奇
地点：南非开普敦

Project Name: Pearl Valley 276
Design Company: Antoni Associates (www.okha.com)
Designer: Mark Rielly, Jon Case, Sarika Jacobs
Photographer: Adam Letch
Location: Cape Town, South Africa

"环翠拥水"设计尽可能选用自然有机材质，如原木、石材等。质感丰富的材质让空间有了更多的家居感觉，倍感温馨。设计在运用水景、壁炉等元素的同时，也利用了很多基本的元素。粗糙、自然质感的材质遍布于整个空间。木质的地板，不仅与天花的混凝土形成对比，也反衬着自然石材的温润、人造石材的精致。

定制的酒窖与客厅毗邻。客厅内，镭射切割的金属滑动板材界定着各个功能分区。开放布局的厨房、非正式的开放酒吧构成了非正式的家庭活动空间。此处空间全部与娱乐休息室相连。介于家庭休息室与 SPA 之间，有一双面壁炉。壁炉外表包裹着石材。

木材、织纹皮革、亚麻制品让奢华变得低调而不张扬。透过黑色的印刷制品、富有生机的面料，可见业主对深色调的喜好。

空间的灯光照明，大胆而又谨慎。富有层次的光照，给人惊叹的同时，调节着不同场合的情绪。幽深之外，到处皆是精致灯光的身影。外面四周的边缝，也因此有了一种温暖的闪亮。空间特别运用了隐式的照明，升华着有机、自然的饰面。餐厅内，明星设计师汤姆·迪克森的照明设计提升着餐厅的魅力。主卧室里的灯光照明同样品牌手笔，荷兰家具品牌 MOOOI 的制品随机摆放。

考究的细部、丰富的饰面营造出一个极富诱惑力与感染力的家居环境。

一层平面图 / First Floor Plan

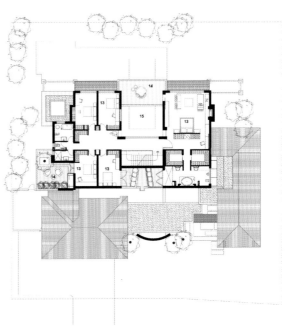

二层平面图 / Second Floor Plan

The design was to focus on the use of natural organic materials such as timber and stone. These tactile materials add a sense of homeliness and warmth to the contemporary architecture. A number of elemental forces are captured in the use of water features and fireplaces that were added to the project. Raw, natural materials were used throughout the project. Timber flooring was used throughout, which contrasts with the raw off-shutter concrete ceilings, warm stone and marble cladding.

Key features of this include the bespoke wine cellar adjacent to the dining room which has been screened off by laser-cut metal sliding panels. The informal family spaces include an open plan kitchen, informal dining and open bar, all linked to the entertainment lounge. A feature marble clad two-sided fireplace divides and screens off the family lounge from the spa room.

The furniture is modern and complementary to the experience of the home. Tactile finishes including timber, textured leathers and raw linens add a sophisticated sense of understated luxury. The clients' love for color has been introduced with injections of bold prints and vibrant fabrics.

A combination of bold and discreet lighting was used to create a "wow" factor and the layering of lighting set various moods. Subtle lighting has been incorporated in all recesses and feature bulkheads to give a warm glow to peripheral edges. Concealed lighting has also been used to highlight and accentuate the organic natural finishes. In the dining room a collection of Tom Dixon's lights add glamour to this space. In the main bedroom a dramatic statement is effected with the random lights from MOOOI.

This house is unique in its detail and quality of finish and reflects a living environment that is seductive and inspired by family living.

All Luxury but Nothing
赤足的奢华

项目名称：珍珠谷之"赤足的奢华"
设计公司：安东尼联营公司
设计师：马克·里利、乔恩·凯斯、萨丽卡·雅各布斯
摄影师：亚当·莱奇
地点：南非开普敦

Project Name: Pearl Valley 334
Design Company: Antoni Associates (www.okha.com)
Designer: Mark Rielly, Jon Case, Sarika Jacobs
Photographer: Adam Letch
Location: Cape Town, South Africa

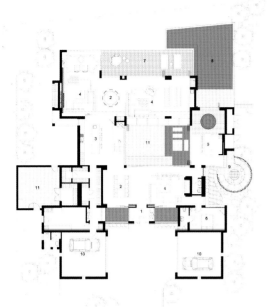

一层平面图 / First Floor Plan

二层平面图 / Second Floor Plan

空间设计以庭院为中心，庭院里有一反射泳池。各生活空间、娱乐区域，全以推拉门、通风门界定，整个空间也因此有了一种通透的感觉。主休息室为双高空间，其天花以构架作为支撑。休息室的中间夹层用作书房。

娱乐区、生活区以台阶相连。厨房面向中央院落，现代、整洁、有序，所有设施组合设计。悬浮的楼梯，以人造石板作为步道，通往上面的卧室区。卧室区由四个套房组成。

水池之上有一栈桥。过栈桥，便可到达酒吧区。酒吧附设的双层SPA，正好凌驾于水池之上。侧墙以盖板作为装饰，美观的同时又可收纳一电视，便于观看电视节目。酒吧外面还有一个区域，在那里可以燃起篝火，尽兴畅谈。

楼上的卧室区延续着同样令人欢愉的质感饰面。卧室的卫生间同样用木板作为四壁的覆盖，盥洗空间也因此多了层层的温暖与低调的奢华。

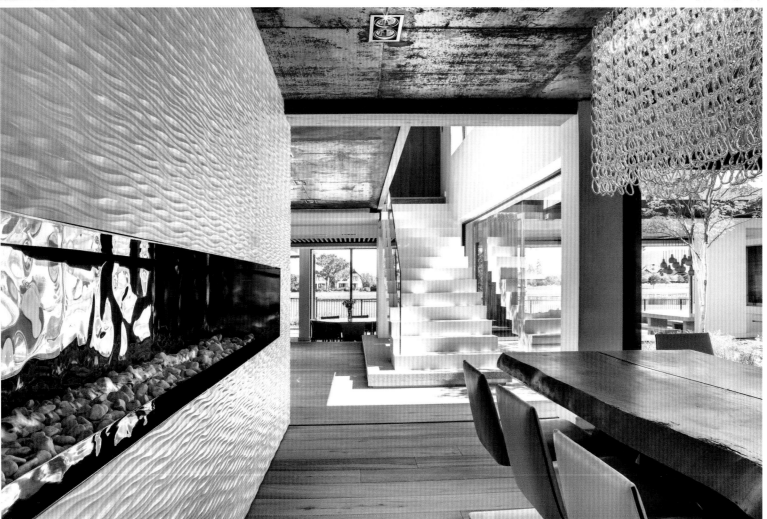

The house has been designed around a central landscaped courtyard with a reflective pond. Sliding and stacking doors open from all the surrounding living and entertainment spaces, and add to the transparency of the house and its natural surrounds. The formal lounge is a large double-volume cathedral-like space with a trussed ceiling which is overlooked by the mezzanine study situated above the formal dining room.

Linking the formal areas to the entertainment rooms that are located a few steps down; the kitchen also looks out onto the central courtyard. The kitchen is modern and has been neatly fitted with integrated appliances and seamless finishes. A floating stair finished in Caesarstone slabs leads up to the sleeping level with four en suite bedrooms.

Floating steps over the reflective pond lead to the bar area, which has a decked spa overlooking the pool. Here the surrounding decking has been wrapped up a side wall which incorporates a TV for watching sporting events. The bar area is also linked to an outdoor "boma", which is a casual enclosure with seating area around a fire-pit.

The same sensory pleasure and use of textured finishes has been used in the upstairs bedrooms. Timber wall cladding has been introduced in the en-suite bathrooms which add layered warmth and understated luxury to the wash rooms.

The Sea Melted into the Sky

坐拥海天一色，收藏天地美景

项目名称：露西亚之家
建筑公司：SAOTA 建筑师事务所
室内设计：安东尼联营公司
结构设计：托尼工程师事务所
摄影：SAOTA 建筑师事务所
面积：1 099 m²

Project Name: La Lucia
Construction Company: SAOTA – Stefan Antoni Olmesdahl Truen Architects
Interior Design: Antoni Associates
Structural Deign: Tony Cooksey Structural Engineers
Photography: SAOTA
Area: 1,099 m²

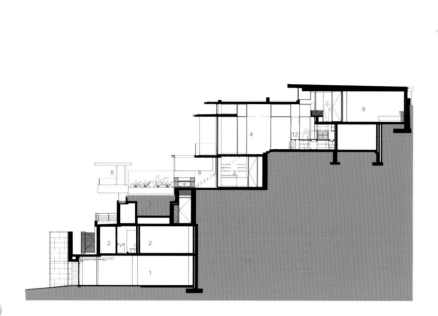

剖面图 / Sectional Drawing

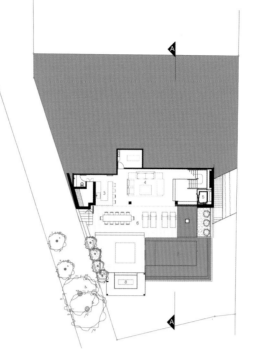

四层平面图 / Fourth Floor Plan

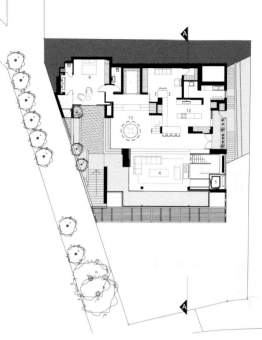

五层平面图 / Fifth Floor Plan

本案力求纳厨房、客厅、餐厅于同一空间之内，在这里不仅有全天候采光，更可尽览美丽的海天风情。次生活区是一个家庭娱乐空间，恰好位于主生活区的下面，便于招待客人的同时，也便于家居生活。泳池露台，既有封闭区域，也有开放区域，满足着不同的休息需要。娱乐休息室位于室外烧烤野餐区附近，内部设有酒吧。泳池平台西部另有凉亭，方便主人欣赏落日余晖。

主、次生活区位于四、五楼。六楼主要用于卧室，包括两个儿童卧室套房和一个小型的儿童休息室。二、三楼主要为客人房、员工区域及私人图书室。一楼除了玄关，还设有5车位的车库。各楼层除了玻璃电梯予以连接，还有一个室外楼梯贯通上下。

因为空间饰面、造型装饰种类繁多，且各具特色，故常常予人极为丰富的空间视觉。内外墙体所用石材，细腻而富有质感，相对而言五楼的客厅、餐厅、厨房，其拱腹所用混凝土则略显粗犷。其他区域，包括各房间所用木质，在延续着如此丰富饰面的同时，又增添了丰富的马赛克饰面。

The design was primarily driven by the need to create a family home, which accommodated the kitchen, living room and dining room in one single space. These areas enjoyed all day sunlight with simultaneously framed views of the sea. The secondary living area was to be a dramatic entertainment space, located on the level immediately below the family level, where the clients could entertain large groups of friends. The pool terrace allows for covered and uncovered areas to relax around the pool. The entertainment lounge accommodates a generous bar, and is close to the outdoor braai area. A dramatic gazebo structure is perched at the Western edge of the pool deck, which allows the owners to enjoy the last hours of the setting sun on the edge of the pool deck.

These 2 living levels are located centrally in the vertical arrangement of the house on the Fourth and Fifth floor. The Sixth floor above accommodates the main bedroom with 2 children's en-suite bedrooms and a small children's lounge. The guest room, a staff area and a private library are located on the second and third story below the entertainment level. The ground floor accommodates the entrance hall and a five car garage. A glass lift connects the building vertically, and an external service stair connects the levels externally.

Visually the building is enriched with the use of a number of interesting finishes and features. These include textured stone cladding to various walls internally and externally. This is contrasted with the roughness of the off-shutter concrete soffit to the living room, dining room and kitchen on the fifth floor. The finishes are rich and varied to various other areas in the house, including timber cladding to various rooms, and richly colored mosaic finishes.

Garden Above City

铺陈在城市上方的空中庭园

项目名称：城市塔
建筑公司：SAOTA 建筑师事务所
建筑师：格雷格·图恩、艾娜·傅立叶
设计公司：OKHA 室内设计
设计师：亚当·考特
摄影师：杜克
地点：南非约翰内斯堡

Project Name: Sandhurst Towers
Construction Company: SAOTA - Stefan Antoni Olmesdahl Truen Architects(www.saota.com)
Architect: Greg Truen, Ina Fourie
Design Company: OKHA Interiors(www.okha.com)
Designer: Adam Court
Photographer: Dook
Location: Johannesburg, South Africa

一层平面图 / First Floor Plan

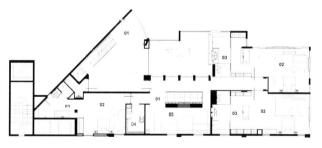

二层平面图 / Second Floor Plan

三层平面图 / Third Floor Plan

"城市塔"，三层顶楼空间。时髦的家居场所，如同21世纪城市生活空间的演绎，内部有个性张扬的装饰，有现代的科技，有定制的灯光，有自动化的设施，空间实用而感性脱俗。

现代奢华的生活空间，借助于本案，有了一种全新诠释。四居室的空间，为生活在城市核心区域的人们提供了一个安静、安全、舒适的场所。城市商业内核，11层高的巅峰空间虽内外之间的界限模糊，但喧嚣早已经无踪影。1 000平方米的超大空间，因花园的存在，真正地成了铺陈在城市上方的空中庭园。

三层的空间，层层得以"格式化"。借助于用材，空间的质感、透明感、反射感得以强化、升华。因此，不同区域之间有了一种统一的流动感。原木、花岗岩等有机材质的使用，为空间平添了一种触感，同时消弥了空间过多的现代气质。建筑、设计、装饰在此实现了三位一体。一切尽为满足业主要求，诺大的空间如同一块面料，根据业主而量身打造。空间里外畅通、合二为一。但各空间依然独立地保持着自己的个性，而不仅仅是传承着、继续着以前的环境。

黑白两色的超大空间，间以其他色彩作为点缀。艺术品、雕塑、玻璃器皿、瓷器及其他装饰铺陈增加了空间的层次感，加深了空间的力度，为空间注入了灵魂与性情。有奢华，但却不浮躁，一切都是那么优雅、低调。灰白、黑色的体块映照着空间的轮廓。开放式的布局也因此有了一种极强的视觉观感与空间张力。

This three-level Sandton penthouse was designed into a product that would showcase the evolution of 21st Century urban living, reflect individualism as well as evolve the technological aspects of the "smart home", with tailored lighting and automation. The brief was both very practical as well as being almost metaphysical and certainly emotive.

This is an establishment of a new standard in modern urban luxury living. An opportunity to illustrate how a family – unit there are four bedrooms – can live in central Sandton in complete calm, security and extreme comfort. Despite the fact that this was an 11th floor penthouse apartment in a commercial city center, an unobstructed indoor/outdoor flow not normally associated with metro living is made. A wall-less 1000m2 penthouse this project makes with a garden, which gives the impression of levitating above the city mass.

Every floor is reformatted, creating a continual flow from zone to zone, utilizing materials that emphasized and exaggerated the sense of space, transparency and reflection. Organic textural elements through the use of timber and flamed granite are introduced to soften and add a counterpoint to the contemporary drama. From the outset the interior architecture and interior design and décor were inextricably linked. In effect the project tailored in every way to the client's needs like a couture garment. A fluid connectivity throughout is maintained but each zone is given its own very specific look, not to have each simply a continuation of the previous environment.

The overall palette throughout the entire 1,000 square meters is black and white with additional color used as an exotic accent. Artwork, sculpture and a large collection of glassware, ceramics and other accessories play a vital role in layering the space and adding soul, depth and character. There are no over-the-top grandiose statements, no extravagant flamboyance but rather a refined and understated luxury. The color palette of off-whites features blocks of black that define silhouettes and contours and add graphic perspective and drama to the open-plan format.

拥山抱海，收藏天地醉蓝美景

Here Is the Hill and the Sea

项目名称: N198
建筑公司: SAOTA 建筑师事务所
建筑师: 格雷格·图恩、斯蒂芬·安
东尼、萨尔斯·普林斯罗
摄影师: 亚当·莱奇
地点: 南非开普敦
面积: 11 000 m²

Project Name: Nettleton 198
Construction Company: SAOTA - Stefan
Antoni Olmesdahl Truen Architects
(www.saota.com)
Architect: Greg Truen, Stefan Antoni,
Jacques Prinsloo
Photographer: Adam Letch
Location: Cape Town,South Africa
Area: 11,000 m²

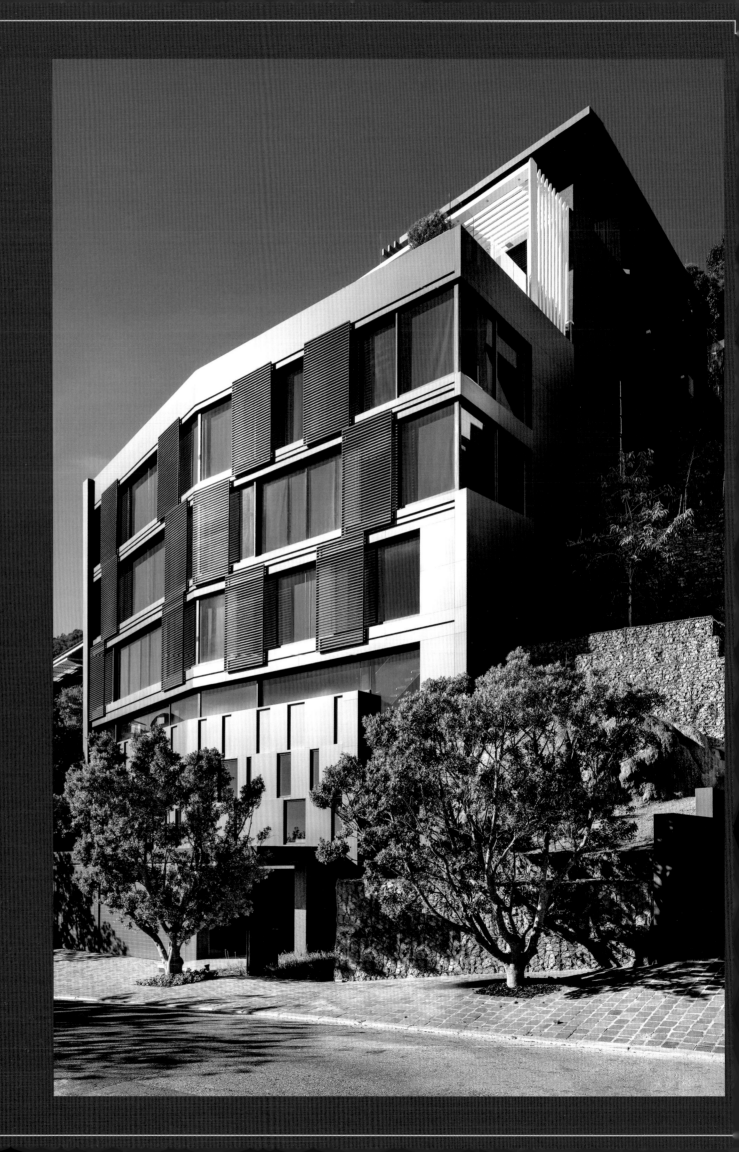

本案的设计灵感源于周围的山景。漆黑色的立面使建筑量体如同隐于山林之中。除了借景狮子头山的壮阔，设计还将海景与日出日落引入空间。

下层空间因屏风、百叶窗的使用显得更加别致而庄重。如此设计，其实也是对太阳直射的应对。客厅东西两面朝向，通透的感觉让此处空间仿如开放的亭台楼阁一般。

饰面与细部极为考究，给人一种整体上的视觉流通感。建筑的外面包裹着铝质的材质，空间外表因此有了一种矍铄、精确的感觉。内部，因核桃木的使用，而给人一种温暖的质感。同时，空间也不乏黑色大理石、玻璃等材质。

家居空间设计别有一种喜庆之感。核桃木包裹的环形玄关，是其主要特色。背光的玻璃板置以梦幻般的照明，放大着空间的视觉感，同时也产生了一种超脱尘世的感觉。悬空的木质楼梯借助于不锈钢的拉杆作为固定，楼梯的栏杆也因此有了一种新鲜的感觉。核桃木板镶嵌的厨房，花岗岩围塑的无边际水池，雕塑质感的悬臂碳化纤维酒吧也构成了空间的特色。

Inspiration of this project was drawn from the mountain and dark colors were used on the façade, allowing the building to visually recede into the mountain instead of being an obtrusive construction. The site enjoys spectacular views, both of the sea and Lions Head and these views and the impact of the sun were key informants contributing to the overall design.

The sun being both a defining and also harsh influence on the property inspired the choice of screens, shutters and louvers that give the lower levels its distinctive gravitas. The living room can open up onto the west and the east completely, giving it the feeling of an open pavilion.

The finishes and detailing have been very carefully considered to achieve an integrated and visually effortless whole. The exterior of the building is clad in powder-coated aluminium which resulted in a very robust and precise surface finish. Internally, a much warmer look was achieved by using walnut timber. Black marble & glass were integrated into the design as accents.

The home is peppered with unusual design "delights". The circular entrance area, clad in walnut timber, is one of the main features of the house. The space is amplified by a fascinating lighting installation of backlit slumped glass, that renders an "otherworldly" effect. An elegant floating timber staircase employs hanging stainless steel rods for a refreshing take on a balustrade. Others include the walnut-clad kitchen box, the granite-clad rim-flow pool and the sculptural cantilevered carbon fiber bar.

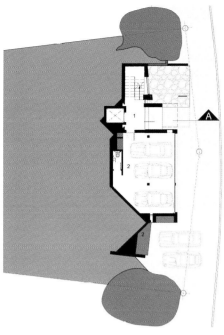

一层平面图 / First Floor Plan

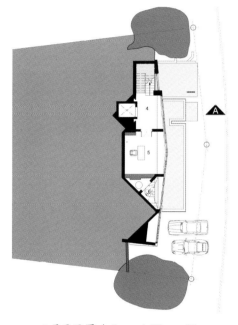

二层平面图 / Second Floor Plan

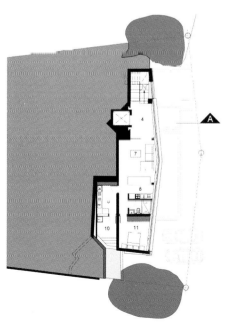

三层平面图 / Third Floor Plan

With Art Collection, Comes Life Interest

收藏艺术，亦收藏生活的无限情趣

项目名称：牛津 49
设计公司：OKHA 室内设计
设计师：亚当·考特，塔维亚·法拉奥
摄影师：埃尔斯·杨
地点：南非约翰内斯堡

Project Name: Oxford 49
Design Company: OKHA Interiors (www.okha.com)
Designer: Adam Court, Tavia Pharaoh
Photographer: Else Young
Location: Johannesburg, South Africa

一层平面图 / First Floor Plan

二层平面图 / Second Floor Plan

本案业主为法国商人，业主对艺术、娱乐情有独钟。五居室空间可谓是一专多能，不仅可以满足家居生活的需要，而且起到招待朋友、接待贵宾的作用。独特的家居氛围中弥漫着一种宁静、整洁与优雅的气氛。花瓶、雕像、艺术品、家具悄然而立，专供宾主观赏。

The JHB home of a French businessman with a love of art & fine entertaining. This 5 bedroomed house was designed to be a multi-functional home for family, visiting friends and executive guests. The brief was to express an individual ambient character but maintain a calm and uncluttered elegance. Every vase, sculpture, artwork and furniture object is given enough meditative space to allow it to be seen and appreciated.

图书在版编目（CIP）数据

家族荣耀　豪宅铭刻 / 黄滢 马勇 主编 . – 武汉 : 华中科技大学出版社 , 2014.9

ISBN 978-7-5680-0423-7

Ⅰ . ①家… Ⅱ . ①黄… ②马… Ⅲ . ①别墅 – 建筑设计 – 世界 – 图集 Ⅳ . ① TU241.1–64

中国版本图书馆 CIP 数据核字（2014）第 224897 号

家族荣耀　豪宅铭刻（上、下）　　　　　　　　　　　　　　　黄滢 马勇 主编

出版发行：华中科技大学出版社（中国·武汉）

地　　址：武汉市武昌珞喻路 1037 号（邮编：430074）

出 版 人：阮海洪

责任编辑：熊纯　　　　　　　　　　　　　　　　　责任监印：张贵君

责任校对：岑千秀　　　　　　　　　　　　　　　　装帧设计：筑美空间

印　　刷：利丰雅高印刷（深圳）有限公司

开　　本：889 mm × 1194 mm　1/12

印　　张：38（上册 20 印张，下册 18 印张）

字　　数：228 千字

版　　次：2015 年 01 月第 1 版 第 1 次印刷

定　　价：598.00 元（USD 119.99）

投稿热线：（020）36218949　　duanyy@hustp.com

本书若有印装质量问题，请向出版社营销中心调换

全国免费服务热线：400-6679-118 竭诚为您服务